Mycorrhizae
AND PLANT HEALTH

Edited by
F. L. Pfleger and R. G. Linderman

APS PRESS

The American Phytopathological Society
St. Paul, Minnesota

This publication is based, in part, on presentations from a symposium entitled "A Reappraisal of Mycorrhizae in Plant Health" held in conjunction with the annual meeting of The American Phytopathological Society, August 12, 1992, in Portland, Oregon. To make the information available in a timely and economical fashion, this book has been reproduced directly from computer-generated copy submitted in final form to APS Press by the editors of this volume. No editing or proofreading has been done by the Press.

Reference in this publication to a trademark, proprietary product, or company name by personnel of the U.S. Department of Agriculture or anyone else is intended for explicit description only and does not imply approval or recommendation to the exclusion of others that may be suitable.

Library of Congress Catalog Card Number: 93-72994
International Standard Book Number: 0-89054-158-2

Printed in the United States of America on acid-free paper

The American Phytopathological Society
3340 Pilot Knob Road
St. Paul, Minnesota 55121-2097, USA

TABLE OF CONTENTS

EFFECTS OF MYCORRHIZAE ON BIOGEOCHEMISTRY AND SOIL STRUCTURE

CURRENT STATUS OF FIELD APPLICATION OF MYCORRHIZAL FUNGI

VAM FUNGAL SYSTEMATICS

GENETIC VARIABILITY IN MYCORRHIZA FORMATION AND FUNCTION

MYCORRHIZAE AND PLANT HEALTH

PREFACE

Most plants on Earth have a symbiotic association in their roots with soil fungi known as mycorrhizae. There is widespread agreement among plant and soil scientists that mycorrhizae are beneficial to the growth and health of plants. However, some arbuscular mycorrhizae (AM) and vesicular arbuscular mycorrhizal fungi here after collectively called VAM in this book, have fungal species and strains within a species that vary greatly in their association with plants ranging from mutualistic to parasitic. Suffice it to say, that most plants form mycorrhizae and that this association is critical to the well-being of plants, especially in natural ecosystems. As D. H. Marx and N. C. Schenck stated 10 years ago in the APS book on Challenging Problems in Plant Health, "on most plants mycorrhizae are as much a natural component of roots as chloroplasts are components of leaves". Most plants have evolved with mycorrhizae and their very existence depends on the mycorrhizal association.

The Mycorrhiza Committee of the American Phytopathological Society proposed that a symposium be held at the 1992 Annual Meeting in Portland, Oregon, where the theme was "Healthy Plants, Healthy Planet". The objective of this symposium was to inform the Society membership at large of the importance of mycorrhizae in plant health beyond the general thinking that mycorrhizae improve plant growth due only to improved P uptake from the soil. A "reappraisal" of the role of mycorrhizae in plant health was the theme of the symposium which focused on mycorrhizae and their role in plant health in diverse ecosystems from mine spoils reclamation sites to agroecosystems where suitability of production of crops is now highly dependent on major inputs of chemical fertilizers and pesticides. Furthermore, we wanted to present evidence for mycorrhiza effects on reducing plant diseases, alleviating soil chemical and physical stresses due to drought and heavy metal pollution, and on improvements in soil structure and stabilization. Application technology is being developed rapidly worldwide on biological diversity and functioning of mycorrhizae, from studies in the field to those at the molecular/biochemical level.

This book is an expanded version of the Symposium proceedings and emphasizes the key role mycorrhizae play in our quest to have healthy plants on Earth and sustainable systems of agricultural and forest production. The contents address the role of mycorrhizae in plant health, including VA mycorrhizae (VAM) that occur on most agricultural and natural ecosystem plants, and ectomycorrhizae that occur on several major families of trees being managed for wood products. We did not address other kinds of mycorrhizae such as ericoid mycorrhizae that occur on most ericaceous plants, or the endomycorrhizae that occur on orchids.

In all ecosystems, mycorrhizae link plants and soil, and that coupling influences most of the dynamics that occur in the mycorrhizosphere. In agroecosystems, current production practices may have damaged or disturbed that coupling, and there is a need to develop an understanding and technology to re-couple plants and soil with mycorrhizae, emulating the balance that occurs in undisturbed ecosystems and returning our crop production systems to a level of sustainability that allows for reduced inputs of pesticides and fertilizers. Additionally, it is our hope that this book will increase awareness and appreciation of mycorrhizae in the thinking and practices that guide our present and future management strategies of agriculture and forestry. Toward that end, we have presented information on the role of VAM and ectomycorrhizae in suppression of plant diseases, including discussion of mechanisms of interaction between mycorrhizae and pathogens (Linderman and Duchesne); the role of mycorrhizae in reclamation of disturbed sites such as from mining activities (Pfleger et al. and Malajczuk et al.); the effects of cultural practices in agricultural and forest nursery systems on mycorrhizae, including cultural practices such as fertilization and pesticide application, weed management, and host genetics (Kurle and Pfleger, and Cordell and Marx); the effects of air pollution on plant health via the mycorrhizal connection (Shafer and Schoenberger); the effects of VAM and ectomycorrhizae on the biogeochemistry of soil regarding nutrient cycling and changes in soil physical structure resulting from aggregation (Miller and Jastrow, and McFall); and on current research activities dealing with practical applications of mycorrhizae in agriculture and forestry (Safir and Castellano) as well as indepth studies on VAM fungal systematics (Bentivenga and Morton) and genetic variability of fungal symbionts in relation to mycorrhiza formation and function (Kropp and Anderson).

The logical result from this endeavor is a more comprehensive view of the important, even critical, role that mycorrhizae have in the health of most plants on Earth. The science of Mycorrhizology is expanding rapidly and will soon make the mycorrhiza connection with other sciences of plant systematics, ecology, and physiology; horticulture; agronomy; soil science; climatology; and certainly plant pathology.

F. L. Pfleger
Dept. of Plant Pathology
University of Minnesota
St. Paul, MN 55108

R. G. Linderman
USDA-ARS
Horticultural Crops
Research Laboratory
Corvallis, OR 97330

CONTRIBUTORS

Anne J. Anderson | Biology Department, Utah State University, Logan, UT 84322-5305

S. P. Bentivenga | Department of Plant Pathology, 401 Brooks Hall, Box 6057, West Virginia University, Morgantown, WV 26506-6057

M. Brundrett | Soil Science and Plant Nutrition Group, University of Western Australia

M. A. Castellano | USDA-Forest Service, Pacific Northwest Research Station, Forestry Sciences Laboratory, Corvallis, OR 97331

C. E. Cordell | Forest Pest Management, Region 8, USDA Forest Service, P. O. Box 2680, Asheville, NC 28802

Luc C. Duchesne | Forestry Canada, Petawawa National Forestry Institute, Chalk River, Ontario, Canada KOJ 1JO

J. D. Jastrow | Environmental Research Division, Argonne National Laboratory, Argonne, IL 60439

Bradley R. Kropp | Biology Department, Utah State University, Logan, UT 84322-5305

J. E. Kurle | Department of Plant Pathology, University of Minnesota, St. Paul, MN 55108

R. G. Linderman | USDA-ARS, Horticultural Crops Research Laboratory, Corvallis, OR 97330

D. H. Marx | Institute of Tree Root Biology, USDA Forest Service, 320 Green Street, Athens, GA 30602-2044

J.S. MacFall
Department of Radiology, Duke University Medical Center, Durham, NC 27710

N. Malajczuk
CSIRO, Division of Forestry, Western Australia

R. M. Miller
Environmental Research Division, Argonne National Laboratory, Argonne, IL 60439

J. B. Morton
Department of Plant Pathology, 401 Brooks Hall, Box 6057, West Virginia University, Morgantown, WV 26506-6057

R. K. Noyd
Department of Plant Pathology, University of Minnesota, St. Paul, MN 55108

F. L. Pfleger
Department of Plant Pathology, University of Minnesota, St. Paul, MN 55108

P. Reddell
CSIRO, Minesite Rehabilitation Research Program, Queensland

G. R. Safir
Department of Botany and Plant Pathology, Michigan State University, East Lansing, MI 48824

M. M. Schoeneberger
USDA-Forest Service, Rocky Mountain Forest and Range Experiment Station, University of Nebraska, Lincoln, NE 68583-0822

S. R. Shafer
USDA-ARS Department of Plant Pathology, North Carolina State University, Raleigh, NC 27695-7616

E. L. Stewart
Department of Plant Pathology, University of Minnesota, St. Paul, MN 55108

Mycorrhizae
AND PLANT HEALTH

ROLE OF VAM FUNGI IN BIOCONTROL

Robert G. Linderman
USDA-ARS Horticultural Crops Research Laboratory
Corvallis, Oregon

Most terrestrial plant species have in their roots a symbiotic association with soil fungi called mycorrhiza. There are several categories of mycorrhizae, of which the largest group, vesicular-arbuscular mycorrhizae (VAM), form with most agricultural crops. VAM fungi exist in soil as thick-walled chlamydospores, or as vegetative propagules in roots, that germinate in the rhizosphere/rhizoplane. Their hyphae penetrate the root cortex, ramifying intercellularly from the point of penetration. The fungus forms special haustoria-like structures (arbuscules or coiled hyphae) within cortical cells, separated from the host cytoplasm by the host plasma membrane and the fungal cell wall. Arbuscules provide increased surface area for metabolic exchanges between the host and fungal partners. VAM fungi also develop extraradical hyphae that grow into the surrounding soil, increasing the potential of the root system for nutrient and water absorption, and contributing greatly to the improvement of soil structure for better aeration and water percolation. New survival spores are usually borne on the extraradical hyphae, although spores of some fungal species are produced primarily intraradically.

Generally, VAM cause few changes in root morphology, but the physiology of the host plant changes significantly. For example, tissue concentrations of growth regulating compounds and other chemical constituents change, photosynthetic rates increase, and the partitioning of photosynthate to shoots and roots changes (16). The improved

1

potential for mineral uptake from the soil accounts for changes in the nutritional status of the host tissues, in turn changing structural and biochemical aspects of root cells. This can alter membrane permeability and thus the quality and quantity of root exudation. Altered exudation induces changes in the composition of microorganisms in the rhizosphere soil, now appropriately called the "mycorrhizosphere" (51,67,82). The net effect of these changes is a healthier plant, better able to withstand environmental stresses (52) and tolerate or reduce the effects of plant diseases.

The purposes of this review are (a) to discuss the role of VAM in the expression of plant disease and the mechanisms involved therein, and (b) to discuss factors that influence the role of VAM in biological control of plant diseases.

VAM EFFECTS ON FUNGAL ROOT PATHOGENS

Since VAM fungi are major components of the rhizosphere, it is logical that they could affect the incidence and severity of root diseases. Whether they do or not has been the subject of numerous reviews over the last 15 years (18,32,46,71,72,73), but there is still controversy. Much of the literature suggests that VAM fungi reduce soilborne disease or the effects of disease caused by fungal pathogens. Dehne (32) reported that disease damage was reduced in 17 out of the 32 reports cited. However, the reports are still mixed, with some indicating no effect of the VAM fungus on disease (6,7,28,84,85), and others indicating increased disease severity (29,30,31). Drawing conclusions is difficult, partly because so many different pathogens and diseases have been involved, and partly because of the experimental conditions of each study. Clearly, one should expect varied results, even if the VAM fungi used had been the same (71).

VAM EFFECTS ON PATHOGENIC ROOT NEMATODES

Root infections by pathogenic nematodes are

generally less severe on VAM plants than on nonVAM plants, but the responses may vary, and the mechanisms involved are controversial (42,44). Symptoms of nematode infection are generally reduced, and often, but not always, nematode populations are reduced (as indicated by number of galls, juveniles or eggs per unit root length)(41). Atilano et al. (3), however, showed an increase in nematode populations on VAM plants. These differences may be due largely to differences between nematode pathogens, but could also be due to differences in VAM fungi and their levels of colonization. Furthermore, as with VAM interactions with fungal pathogens, timing of VAM formation is important.

A number of mechanisms of interaction between VAM and nematode pathogens have been considered, and the evidence supporting each is reasonable. All the proposed mechanisms depend on VAM-mediated changes in host physiology. Changes in root exudation by VAM plants may change the attractiveness of roots to nematode pathogens. VAM may improve host plant vigor, and thus reduce yield losses caused by nematode infection, especially in low P soils and if VAM are established early in the growth cycle, before nematode infection. This mechanism of enhanced P nutrition in the reduced nematode response has been challenged, however, by work that showed no effect from adding P (55). Cooper and Grandison (24,25) eliminated P as a variable by using P-tolerant VAM fungi on plants grown under high P conditions. Still, VAM reduced nematode effects, leading these workers to conclude that increased resistance to nematodes was not entirely due to improved host nutrition, but must involve some other physiological changes in the roots. Studies by Strobel et al. (78) and Oliveira and Zambolim (61) using split root techniques indicated that reduced nematode infection on VAM plants only occurred if the two were together on the same roots. These results suggested that competition for food or space was involved, a mechanism supported by results of Saleh and Sikora (69), Suresh et al. (79), Sitaramaiah and Sikora (75), and MacGuidwin et al. (55). These studies indicated that nematode size was reduced and rate of development of infection was slower in VAM roots than nonVAM roots. Physiological changes in VAM roots could also change resistance to nematodes by increased production of inhibitory substances (79), or by changes in root exudation which could alter mycorrhizosphere populations and affect nematode populations and survival.

While the mechanisms are still controversial, the evidence strongly indicates that VAM suppress nematode infections of roots or reduce nematode effects on plant growth and yield. Undoubtedly, the effects and the mechanisms involved depend on the conditions of the test, the host plant, edaphic conditions, and the species of VAM fungus involved. Nonetheless, it seems safe to say that VAM do play a role in the biological control of root nematode diseases.

VAM EFFECTS ON BACTERIAL DISEASES

The effects of VAM on bacterial diseases have not been explored to any great extent. However, Garcia-Garrido and Ocampo (36) showed that VAM tomato (*Lycopersicon* Miller) plants exhibited greater growth than nonVAM plants inoculated with *Pseudomonas syringae* pv. *syringae* van Hall when the pathogen was added to the foliage three weeks after the VAM fungus. Populations of the pathogen were lower in VAM than nonVAM plants.

VAM EFFECTS ON VIRUS AND FOLIAGE DISEASES

While soilborne diseases caused by fungal, nematode, and bacterial pathogens most often are reduced by VAM, those caused by viral and other foliage pathogens are generally increased in VAM plants (32,46,73). Reports indicate that disease incidence, but not necessarily severity of effect on the plant, is increased in VAM compared to nonVAM plants. Viruses apparently multiply faster in VAM than nonVAM plants. One might suspect that VAM fungi could acquire and vector viruses from root to root between plants, but that apparently does not happen (45). VAM effects on viruses occur throughout the

plant due to changes in the host physiology. As Dehne
(32) pointed out, foliage diseases caused by both
obligate and non-obligate leaf pathogens can be
increased on VAM compared to nonVAM plants, likely due
to enhanced development of the pathogens rather than
to increased incidence or frequency of infections.
That effect was correlated with improved nutrition and
higher physiological activities in VAM than nonVAM
plants (32,73).

MECHANISMS OF VAM EFFECTS ON PLANT DISEASE

Since VAM have such a significant effect on host
plant physiology and on biological interactions in the
rhizosphere soil, it follows that they could affect
the incidence and severity of plant diseases. The
role played by VAM in the biological control of plant
diseases has been the subject of several reviews
(8,18,32,41,46,70,71,72,73), but mixed responses and
interpretations have precluded any clear conclusion
that VAM always suppress plant diseases. Such
inconsistencies should be expected, however,
considering the diverse experimental approaches and
use of different VAM fungi on different hosts in
different soils (71). Part of the controversy also
revolves around the mechanisms of interaction between
VAM and plant pathogens. In reviews of biological
control of plant diseases, mycorrhizae are thought to
contribute to biological control of plant diseases
primarily by means of stress reduction
(10,12,23,62,72). The literature of recent years,
however, indicates that biological control of plant
diseases may be strongly influenced by VAM by one or
more mechanisms, including: (a) enhanced nutrition,
(b)·competition for host photosynthate and infection
sites, (c) morphological changes in roots and root
tissues, (d) changes in chemical constituents of plant
tissues, (e) reduction of abiotic stresses, and (f)
microbial changes in the mycorrhizosphere.

Enhanced Nutrition
The most obvious VAM contribution to reduced root
disease is to increase nutrient uptake, particularly P

and other minerals, resulting in more vigorous plants better able to resist or tolerate root disease. The evidence to support the enhanced nutrition idea comes from experiments where effects comparable to VAM effects were observed when more fertilizer P was added. Davis (28) showed this type of response in his studies on Thielaviopsis root rot of citrus where VAM plants were larger than nonVAM plants unless the latter were fertilized with additional P. Graham and Menge (38) suggested a similar effect, where VAM or added P reduced wheat take-all disease (*Gaeumannomyces graminis* var. *tritici* (Sacc.) Arx & Oliv.), and speculated that enhanced P status of the plants caused a decrease in root exudates used by the pathogen for spore germination and infection. They did not, however, demonstrate increased pathogen spore germination with those treatments. In some cases, reports indicate that VAM or added P increased disease incidence, as in the case of Verticillium wilt (*Verticillium dahliae* Kleb.) of cotton (*Gossypium rirsutum* L.) (31).

In an attempt to clarify the confusion about the role of enhanced P nutrition associated with VAM and root disease expression, Graham and Egel (37) found no differences between Phytophthora root rot levels on VAM and nonVAM citrus seedlings fertilized to be of equal size and P content. Caron et al. (20,21,22) compared responses between VAM and nonVAM tomato plants with a relatively low P threshold requirement to root and crown rot disease caused by *Fusarium oxysporum* f. sp. *radicis-lycopersici* Jarvis & Shoemaker. Added P did not reduce disease response and pathogen populations in rhizosphere soil of nonVAM plants, but did with VAM plants, even though plant growth and tissue P were not different in the two treatments. This work suggests the involvement of some mechanism of disease suppression other than enhanced P uptake. Whether or not enhanced P uptake by VAM is involved either directly or indirectly in disease expression remains controversial. The possibility that improved uptake of other mineral elements from soil could be involved has not been explored.

Competition for Nutrients and Infection Site

Even though VAM fungi depend on the host plant for carbohydrates from photosynthesis, it is not clear whether they compete with root pathogens for nutrients. Dehne (32) indicated that fungal root pathogens could occupy root cortical cells adjacent to those colonized by VAM fungi, indicating a lack of competition. It has been suggested that nematode pathogens, on the other hand, require host nutrients for reproduction and development, and direct competition with VAM fungi has been hypothesized as a mechanism of their inhibition (32,76). There is little or no direct evidence to support that hypothesis, however. On the contrary, VAM appear to more than make up for their nutrient needs by enhancing photosynthesis without limiting the nutrient supply to root pathogens.

Morphological Changes

Localized morphological effects have been shown to occur in VAM roots. For example, Dehne and Schonbeck (33) showed increased lignification of tomato and cucumber (*Cucumis satvus* L.) root cells of the endodermis in VAM plants, and speculated that such responses accounted for reduced Fusarium wilt (*F. oxysporum* f. sp. *lycopersici* (Sacc.) Snyder & Hans.). Becker (15) showed a similar effect on pink root of onion (*Pyrenochaeta terrestris* (Hansen) Goreng, Walker & Larson). Wick and Moore (83) also showed increased wound-barrier formation that inhibited Thielaviopsis black root rot (*Thielaviopsis basicola* (Berk. & Br.) Ferr.) of VAM holly (*Ilex crenata* Thundb.) plants. These few examples indicate that VAM may induce morphological changes in root tissues. In most other studies, however, roots were not examined for anatomical changes, so it remains unknown just how extensively such a mechanism may be involved.

Changes in Chemical Constituents of Plant Tissues

Physiological changes also have been reported to be involved in localized effects on root pathogens. Dehne et al. (34) demonstrated increased concentrations of anti-fungal chitinase in VAM roots, and they also suggested that increased arginine accumulation in VAM roots suppressed *Thielaviopsis* sporulation, a mechanism previously suggested by Baltruschat and Schonbeck (13). Morandi et al. (58) found increased concentrations of phytoalexin-like isoflavonoid compounds in VAM vs. nonVAM soybeans *(Glycine max* L.). They postulated that such materials could account for increased resistance to fungal and nematode root pathogens, compared to nonVAM plants. More recently, however, Morris and Ward (59) reported chemoattraction of pathogen zoospores by isoflavonoids from soybeans. It would appear that such compounds, as well as other compounds, could have different effects on disease incidence and/or severity. Just what role VAM play in these processes remains unknown, lacking direct evidence.

Alleviation of Abiotic Stress

Environmental stresses influence the incidence and severity of biotic plant diseases and cause some abiotic diseases. VAM increase plant tolerance to such stresses by various mechanisms. In this context, VAM can function to biologically reduce plant diseases by virtue of their capacity to reduce effects of predisposing stress factors such as nutritional stress (deficiency or excess), soil drought, and soil toxicities.

Because of the greater volume of soil explored by extraradical hyphae of VAM fungi compared to nonVAM roots, nutrient mineral elements that are relatively unavailable because they are bound to soil particles (i.e. P, Cu, and Zn) are absorbed by VAM fungal hyphae and translocated to the root from beyond the zone of nutrient depletion around the root. VAM are able to acquire these and relatively mobile nutrients like nitrogen (N) from soils where deficiency levels would otherwise create plant nutrient stress. Nutrient stress may weaken the plant, making it more susceptible to pathogen ingress, or more sensitive to other environmental stresses such as cold or heat. Thus, VAM contribute greatly to the general health of plants by helping to avoid nutrient stress and associated disease-predisposing effects.

Drought stress is another abiotic factor that predisposes plants to attack by some opportunistic pathogens. Extraradical hyphae of VAM fungi may absorb water under soil drought conditions (39), and thus help plants to tolerate drought. There is, however, controversy over whether direct absorption of water from soil is the mechanism whereby VAM help plants tolerate drought, or whether the increased P uptake by VAM is responsible (60). Others suggest that VAM change the physiology of plants in ways that make them more drought tolerant than nonVAM plants (4,5,26,27).

VAM plants are less sensitive than nonVAM plants to soil toxicities resulting from excess salts (40, 65) or mineral elements such as heavy metals (16). There is controversy about the mechanisms involved, with some work implicating improved P nutrition in salt tolerance (66), but little evidence exists as to the mechanisms of heavy metal tolerance. In both cases, however, the toxic materials are selectively excluded or are somehow altered to prevent toxic effects on plant growth. As with other stresses, VAM plants are more tolerant to soil toxicities and thus maintain a higher level of growth and health than nonVAM plants, and thus may be less susceptible to diseases.

Microbial Changes in the Mycorrhizosphere

While any of the above mechanisms, or combinations thereof, could be involved in VAM suppression of root diseases, one that should be considered more carefully is the VAM alteration of rhizosphere populations of antagonists. Even though the evidence is clear that microbial shifts do occur in the mycorrhizosphere (57,74), most studies have not considered those changes relative to biological control of diseases, so relatively little data are available to support such a mechanism.

The concept of the "mycorrhizosphere" implies that mycorrhizae significantly influence the microflora of the rhizosphere by altering root physiology and exudation. In addition, extraradical hyphae of VAM fungi provide a physical or nutritional substrate for bacteria. Analysis of rhizosphere soil of VAM and nonVAM plants in several studies (2,9,57,74,82) indicated both qualitative and quantitative changes in the mycorrhizosphere soil of VAM plants, compared to rhizosphere soil of nonVAM plants. These microbial shifts were clearly time-dependent and dynamic, changing as the plants developed. Meyer and Linderman (57) used selective media to demonstrate differences in populations of taxonomic and functional groups of bacteria in the rhizosphere and rhizoplane of VAM and nonVAM plants. Similarly, populations of bacteria and actinomycetes in pot cultures of different VAM fungi were quantitatively and qualitatively analyzed by Secilia and Bagyaraj (74). They showed that total populations of bacteria in the mycorrhizosphere soil of VAM plants were greater than in the rhizosphere soil of nonVAM plants. Effects of VAM on other microbial groups, including nitrogen fixing bacteria, actinomycetes, and morphological and physiological groups of bacteria (Gram positive and negative bacteria, spore formers, urea hydrolyzers, and starch hydrolyzers) varied with each VAM fungal species. Furthermore, urea hydrolyzers were present in pot-culture soil of all the VAM plants, but were absent in soil from the nonVAM plants. Vancura et al. (82) documented the selective effects of VAM fungal extraradical hyphae on bacteria from within the mycorrhizosphere. They did not, however, evaluate the antagonistic potential of the microbes associated with the hyphae. These studies demonstrate that VAM influence the microbial populations in the mycorrhizosphere soil; many of those microbial shifts could influence the growth and health of plants.

VAM fungal symbionts produce extraradical hyphae that may extend several centimeters out into the soil and exude organic materials that are substrates for other soil microbes. These hypha-associated microbes frequently produce sticky materials that cause soil particles to adhere, creating small aggregates that impart structure to soil, allowing for improved aeration, water percolation, and stability (81). Forster and Nicolson (35) analyzed the microbial

composition of such aggregates, and identified a range
of fungi, bacteria, actinomycetes and algae, including
cyanobacteria. These microbial associates of VAM
fungi may profoundly affect the further development of
hyphae in soil, and their metabolic products could be
absorbed by the hyphae and translocated to the host.
The specific functional composition of these
aggregates, and the metabolic products produced
therein, are virtually unknown.

VAM formation alters the selective pressure on
populations of soil microorganisms, some of which can
antagonize root pathogens. For example, sporangium
and zoospore production by the root pathogen
Phytophthora cinnamomi Rand. was reduced in the
presence of rhizosphere leachates from VAM plants,
when compared to leachates from nonVAM plants (57).
Similarly, more actinomycetes antagonistic to selected
fungal and bacterial pathogens were isolated from VAM
pot culture plants than from nonVAM plants grown under
the same conditions (74); numbers of antagonists
varied among pot cultures of different VAM fungal
species.

Other studies have indicated that disease
suppression by VAM involved changes in
mycorrhizosphere microbial populations. The work of
Caron et al. (19,20,21,22) indicated a reduction in
Fusarium populations in the mycorrhizosphere soil of
tomatoes and a corresponding reduction in
root rot in VAM plants relative to nonVAM plants,
possibly due to increased antagonism in the VAM
mycorrhizosphere. Their studies also showed that
reduced disease incidence was independent of the level
of P nutrition, but dependent on the nature of the
growth substrate. Another study, by Bartschi et al.
(14), indicated protection of host roots against *P.
cinnamomi* root rot when plants were pre-inoculated
with a mixture of VAM pot culture inocula. The
authors concluded that a mixture of VAM fungi was more
effective than single fungi, but effects could also
have been due to buildup of antagonists in the pot
cultures, as demonstrated by Secilia and Bagyaraj
(74).

These results indicate that VAM fungi are

relatively tolerant of antagonists that inhibit fungal pathogens by one or more mechanisms. They further suggest that VAM fungi, which evolved with plants, are highly rhizosphere-competent and are compatible with such antagonists and even function in concert with them (51). The possibility that antagonistic rhizosphere bacteria or fungi might inhibit mycorrhizal fungi and thereby reduce their effectiveness was tested by Krishna et al. (48), who observed that the pathogen antagonist *Streptomyces cinnamomeous* reduced sporulation and colonization of *G. fasciculatum* (Thaxter) Gerd. & Trappe emend. Walker & Koske on finger millet (*Panicum* L.) if it was added two weeks before the VAM fungus. In spite of that response, however, the combination of the two organisms resulted in greater plant dry weights than if either was used alone.

In extensive trials evaluating interactions between VAM fungi and many fungal or bacterial antagonists, Linderman et al. (54) found little or no adverse effects of bacterial and fungal biocontrol agents on establishment and function of VAM on onion (*Allium cepa* L.). Other studies (63,64) showed the lack of adverse effects or even stimulation of VAM fungi (17) by biocontrol agents, whether applied as seed treatments or added to the soil. Earlier, Meyer and Linderman (56) had shown a positive interaction between the antagonist and plant growth-promoting rhizobacterium *Pseudomonas putida* (Migula) and VAM on subclover (*Trifolium subterraneum* L.). Such interactions must also occur to varying degrees in the rhizospheres of plants grown in pathogen-infested soil, although such evaluations are rarely conducted.

Changes in mycorrhizosphere populations of antagonists to pathogens seems a likely explanation for many of the reported effects of VAM on diseases. Yet, with the exception of those reports mentioned above, few workers have considered that mechanism. Selective increases in numbers of antagonists in the mycorrhizosphere are possible only if the antagonists are present in the background soil or growth medium. Thus, if potentially effective antagonists are present and are increased by VAM, then disease suppression could be expected. On the other hand, if potential antagonists were not present in the soil, then no effect of VAM might result. Reviews on effects of VAM on disease (32,72) would indicate that VAM can increase, decrease, or have no effect on disease. Such responses could indicate that effects on disease correspond to VAM increases in mycorrhizosphere populations of antagonists or deleterious microbes that could enhance disease (80). This hypothesis, further complicated by the differential effects of strains or species of VAM fungi associating with different host genotypes, has not been tested. However, such complex interactions could explain the inconsistencies between studies using different hosts, VAM fungi, and soils from various parts of the world.

FACTORS INFLUENCING MANAGEMENT OF VAM IN BIOCONTROL

Timing and Extent of VAM Formation

When VAM are reported to suppress root disease, they generally must be established and functioning before invasion by the pathogen. This has been demonstrated by Stewart and Pfleger (77) on Pythium and Rhizoctonia root rot of poinsettia (*Euphorbia pulcherrhima* Willd. ex Klotzsch), by Bartschi et al. (14) on Phytophthora root rot of Lawson cypress (*Chamaecyparis lawsoniana* (A. Murr.)Parl.), and by Rosendahl (68) with Aphanomyces root rot of pea (*Pisum sativum* L.). That this would be the case seems logical considering both the faster infection rate of most fungal root pathogens, compared to VAM fungi, and the time needed for VAM effects on the host physiology to occur. Furthermore, other reports have indicated that established root infections by various pathogens can reduce colonization by VAM fungi and therefore the potential for positive effects on disease incidence or severity (1,6,68,84,85). Sometimes, however, root pathogens and VAM fungi occupy adjacent tissues in roots without apparent effects on each other (21,32,73).

If early VAM formation is required for root disease suppression, then what processes are involved? No one has demonstrated direct interactions between

VAM fungi and pathogens, so indirect effects on host morphology and/or physiology or mycorrhizosphere microbial shifts must be involved. However, physiological effects could be localized or systemic. Aphanomyces root rot of peas was only suppressed by VAM when the two organisms were present on the same roots (68). A similar response occurred with Phytophthora root rot of citrus (*Citrus sinensis* L.) in a split root study by Davis and Menge (29), leading them to conclude that the effect was not systemic. However, Rosendahl (68) showed that oospore production was reduced on nonVAM pea roots split from VAM roots, compared to plants with no VAM on either root system. Similarly, Davis and Menge (29) showed that citrus roots opposite the split from VAM roots had less Phytophthora root rot than if split from nonVAM roots. These two studies suggest that VAM effects could be both localized and systemic, probably involving two separate mechanisms. The systemic effect could have been a P effect, while the localized effect was due to some morphological or physiological change in the root tissues in the immediate vicinity.

It is important to consider the time or frequency of observations within a disease cycle in evaluating effects of VAM on diseases. Most studies have evaluated the interaction only once, usually at the end of the experiment. Caron et al. (21) demonstrated that the interaction of VAM and Fusarium root and crown rot of tomato was constant throughout a 12-week experiment, but the percent root necrosis was only significantly reduced at 3 of the 12 observations. Had measurements been made at other than those times, reported effects would have been different.

Inoculum Level of the Pathogen
The potential for biological control to occur in any production system is directly related to the inoculum potential of the pathogen. A high pathogen inoculum density can overwhelm biocontrol agents (11, 23), and this has been shown in VAM studies as well (47,71). It is difficult to draw conclusions about the potential for biocontrol to occur unless a range of pathogen inoculum densities is used.

Variation in VAM Fungi, Host Genotype, and Chemical and Microbial Composition of Soil

It is biologically fundamental to expect different interactions to occur between different VAM fungi, host plants, and plant pathogens. Relatively few studies have compared a range of VAM fungi for any effect (43), and none has done so regarding effects on plant diseases. Some studies have compared different host genotypes for VAM formation and effect (49), but such studies have not included a pathogen in the interactions. Add to that complex host-VAM-pathogen interaction the variation in soil chemical and microbial composition, and it is no wonder that interpretations and comparisons between different studies are difficult if not impossible. Nevertheless, it is conceivable that the right combination of factors does exist, and the best strategy for exploiting VAM for disease suppression can be found.

VAM Management Strategies

If VAM contribute to disease suppression, then agricultural practices that reduce populations of VAM fungi and associated antagonists could result in increased disease incidence which in turn could result in increased need for and use of fungicides. To assure that compatible combinations of VAM and antagonists occur in the same soil or growth medium, growers could inoculate seeds or transplants to guarantee early establishment on the roots prior to pathogen invasion (51,53). Where large acreages are involved and economic margins thin, altering cultural practices to favor effective VAM fungal species would seem appropriate. In those systems where VAM fungi and associated antagonists could be co-inoculated, as in horticultural crops, management of the microbial composition of the rhizosphere seems plausible (50). Optimum combinations of VAM and antagonists need to be selected and inoculated as early as possible in the production cycle at a time when their delivery to the entire root system is guaranteed. By this means, plants can benefit from both types of organisms in protecting against invasion by pathogens.

CONCLUSIONS

With few exceptions, crop plants have VAM, but the degree of root colonization by VAM fungi and the effects of the symbiosis may vary, depending on the total interaction between host, symbiont, and environment. In most cases, VAM significantly change the physiology and chemical constituents of the host, the pattern of root exudation, and the microbial composition of mycorrhizosphere soil. These changes can greatly influence the growth and health of plants, in part due to the biological suppression of plant diseases. Disease suppression may be the result of reduction of environmental stresses that may limit plant growth and predispose the plants to infection by opportunistic pathogens. More important, however, are the specific morphological and physiological changes that directly or indirectly result in reduced incidence and/or severity of plant diseases in VAM plants compared to nonVAM plants. Experiments have differed in design, the VAM fungal symbiont used, pathogen types and inoculum levels, and the plant growth system used. This variation prevents easy conclusions about the predictability of VAM effects on plant diseases. Where disease reduction is reported, one or more mechanisms can be involved, although the tendency is to implicate only one. Because VAM effects on plant nutrition, especially P uptake, are often so striking, many reports implicate improved P nutrition as a mechanism of disease control. Enough reports on VAM suppression of disease where P effects were excluded are now available to suggest the involvement of other mechanisms. Generally, most studies have not investigated other mechanisms such as morphological changes, changes in disease-suppressing chemical constituents in plant tissues, and changes in rhizosphere populations of antagonistic microbes induced by VAM formation. The mechanisms involved probably are multiple, and will depend on the conditions of the test. The major effect of VAM may be improved nutrition, but secondary effects induced thereafter may contribute significantly to observed effects on disease. Mycorrhizosphere changes in populations of antagonists to specific pathogens depend on having those antagonists present in the background soil. If antagonists are absent and deleterious microbes are present in significant numbers and enhanced by VAM, then disease incidence or severity can be increased. In managing rhizosphere populations to result in biological control of plant diseases, compatible VAM fungi and effective antagonists should be delivered to the production system to guarantee their dominance. Such a management strategy could result in stable biological control of diseases and improve overall plant health.

LITERATURE CITED

1. Afek, U., and Menge, J. A. 1990. Effect of *Pythium ultimum* and metalaxyl treatments on root length and mycorrhizal colonization of cotton, onion, and pepper. Plant Dis. 74:117-120.
2. Ames, R. N., Reid, C. P. P., and Ingham, E. R. 1984. Rhizosphere bacterial population responses to root colonization by a vesicular-arbuscular mycorrhizal fungus. New Phytol. 96: 555-563.
3. Atilano, R. A., Menge, J. A., and Van Gundy, S. D. 1981. Interaction between *Meloidogyne arenaria* and *Glomus fasciculatus* in grape. J. Nematol. 13: 52-57.
4. Auge, R. M., Schekel, K. A., and Wample, R. L. 1986. Osmotic adjustment in leaves of VA mycorrhizal and nonmycorrhizal rose plants in response to drought stress. Plant Physiol. 82:765-770.
5. Auge, R. M., Schekel, K. A., and Wample, R. L.. 1986. Greater leaf conductance of well-watered VA mycorrhizal rose plants is not related to phosphorus nutrition. New Phytol. 103:107-116.
6. Baath, E., and Hayman, D. S. 1983. Plant growth responses to vesicular-arbuscular mycorrhiza. XIV. Interactions with *Verticillium* wilt on tomato plants. New Phytol. 95: 419-426.
7. Baath, E., and Hayman, D. S. 1984. No effect of VA mycorrhiza on red core disease of strawberry. Trans. Br. Mycol. Soc. 82: 534-536.

8. Bagyaraj, D. J. 1984. Biological interactions with VA mycorrhizal fungi. Pages 131-153 in: VA Mycorrhiza, C.L.Powell and D.J. Bagyaraj, eds., CRC Press, Inc. Boca Raton, FL. 234 pp.

9. Bagyaraj, D. J., and Menge, J. A. 1978. Interactions between a VA mycorrhiza and *Azotobacter* and their effects on rhizosphere microflora and plant growth. New Phytol. 80: 567-573.

10. Baker, K. F. 1987. Evolving concepts of biological control of plant pathogens. Annu. Rev. Phytopathol. 25:67-85.

11. Baker, K. F., and Cook, R. J. 1974. Biological control of plant pathogens. W. H. Freeman, San Francisco, CA.

12. Baker, R. 1986. Biological control: an overview. Can. J. Plant Pathol. 8:218-221.

13. Baltruschat, H., and Schoenbeck, F. 1975. Studies on the influence of endotrophic mycorrhiza on the infection of tobacco by *Thielaviopsis basicola*. Phytopath. Z. 84:172-188.

14. Bartschi, H., Gianinazzi-Pearson, V., and Vegh, I. 1981. Vesicular-arbuscular mycorrhiza formation and root rot disease (*Phytophthora cinnamomi*) development in *Chamaecyparis lawsoniana*. Phytopath. Z. 102: 213-218.

15. Becker, W. N. 1976. Quantification of onion vesicular-arbuscular mycorrhizae and their resistance to *Pyrenochaeta terrestris*. Ph.D. Diss., University of Illinois, Urbana. (Diss. Abstr. 76:24041)

16. Bethlenfalvay, G. J. 1992. Mycorrhizae and Crop Productivity. Pages 1-27 in: Mycorrhizae in Sustainable Agriculture, G. J. Bethlenfalvay and R. G. Linderman, eds., ASA Spec. Publ. No. 54., Amer. Soc. Agronomy Press, Madison, WI.

17. Calvet, C., Pera, J., and Barea, J. M. 1989. Interactions of *Trichoderma* spp. with *Glomus mosseae* and two wilt pathogenic fungi. Agric., Ecosystems Environ. 29:59-65.

18. Caron, M. 1989. Potential use of mycorrhizae in control of soilborne diseases. Can. J. Plant Pathol. 11:177-179.

19. Caron, M., Fortin, J. A., and Richard, C. 1985. Influence of substrate on the interaction of *Glomus intraradices* and *Fusarium oxysporum* f. sp. *radicis-lycopersici* on tomatoes. Plant Soil 87: 233-239.
20. Caron, M., Fortin, J. A., and Richard, C. 1986. Effect of phosphorus concentration and *Glomus intraradices* on Fusarium crown and root rot of tomatoes. Phytopathology 76: 942-946.
21. Caron, M., Fortin, J. A., and Richard, C. 1986. Effect of *Glomus intraradices* on infection by *Fusarium oxysporum* f. sp. *radicis-lycopersici* in tomatoes over a 12-week period. Can. J. Bot. 64: 552-556.
22. Caron, M., Richard, C., and Fortin, J. A. 1986. Effect of preinfestation of the soil by a vesicular-arbuscular mycorrhizal fungus, *Glomus intraradices*, on Fusarium crown and root rot of tomatoes. Phytoprotection 67: 15-19.
23. Cook, R. J., and Baker, K. F. 1983. The nature and practice of biological control of plant pathogens, APS Press, St. Paul, MN.
24. Cooper, K. M., and Grandison, G. S. 1986. Interaction of vesicular-arbuscular mycorrhizal fungi and root-knot nematode on cultivars of tomato and white clover susceptible to *Meloidogyne hapla*. Ann. Appl. Biol. 108: 555-565.
25. Cooper, K. M., and Grandison, G. S. 1987. Effects of vesicular-arbuscular mycorrhizal fungi on infection of tamarillo (*Cyphomandra betacea*) by *Meloidogyne incognita* in fumigated soil. Plant Dis. 71: 1101-1106.
26. Davies, F. T., Jr., Potter, J. R., and Linderman, R. G. 1992. Mycorrhiza and repeated drought exposure affect drought resistance and extraradical hyphae development of pepper plants independent of plant size and nutrient content. J. Plant Physiol. 139:289-294.
27. Davies, F. T., Jr., Potter, J. R., and Linderman, R. G. 1993. Drought resistance of mycorrhizal pepper plants independent of leaf P concentration - response in gas exchange and water relations.

Physiol. Plant. 87:45-53.
28. Davis, R. M. 1980. Influence of *Glomus fasciculatus* on *Thielaviopsis basicola* root rot of citrus. Plant Dis. 64: 839-840.
29. Davis, R. M., and Menge, J. A. 1980. Influence of *Glomus fasciculatus* and soil phosphorus on Phytophthora root rot of citrus. Phytopathology 70: 447-452.
30. Davis, R. M., and Menge, J. A. 1981. *Phytophthora parasitica* inoculation and intensity of vesicular-arbuscular mycorrhizae in citrus. New Phytol. 87: 705-715.
31. Davis, R. M., Menge, J. A., and Erwin, D. C. 1979. Influence of *Glomus fasciculatus* and soil phosphorus on Verticillium wilt of cotton. Phytopathology 69: 453-456.
32. Dehne, H. W. 1982. Interactions between vesicular-arbuscular mycorrhizal fungi and plant pathogens. Phytopathology 72: 1115-1119.
33. Dehne, H. W., and Schonbeck, F. 1979. Untersuchungen zum Einfluss der endotrophen Mykorrhiza auf Pflanzenkrankheiten. II. Phenolstoffwechsel und Lignifizierung. (The influence of endotrophic mycorrhiza on plant diseases. II. Phenolmetabolism and lignification.) Phytopath. Z. 95: 210-216.
34. Dehne, H. W., Schonbeck, F., and Baltruschat, H. 1978. Untersuchungen zum Einfluss der endotrophen Mykorrhiza auf Pflanzenkrankheiten. 3. Chitinase-aktivitat und Ornithinzyklus. (The influence of endotrophic mycorrhiza on plant diseases. 3. Chitinase-activity and ornithine-cycle). Z. Pflkrankh. 85: 666-678.
35. Forster, S. M., and Nicolson, T. H. 1981. Aggregation of sand from a maritime embryo sand dune by microorganisms and higher plants. Soil Biol. Biochem. 13:199-203.
36. Garcia-Garrido, J. M., and Ocampo, J. A. 1989. Effect of VA mycorrhizal infection of tomato on damage caused by *Pseudomonas syringae*. Soil Biol. Biochem. 21:165-167.
37. Graham, J. H., and Egel, D. S. 1988. Phytophthora root rot development on mycorrhizal and

phosphorus-fertilized nonmycorrhizal sweet orange seedlings. Plant Dis. 72: 611-614.

38. Graham, J. H., and Menge, J. A. 1982. Influence of vesicular-arbuscular mycorrhizae and soil phosphorus on take-all disease of wheat. Phytopathology 72: 95-98.

39. Hardie, K. 1985. The effect of removal of extraradical hyphae on water uptake by vesicular-arbuscular mycorrhizal plants. New Phytol. 101:677-684.

40. Hirrel, M. C., and Gerdemann, J. W. 1980. Improved growth of onion and bell pepper in saline soils by two vesicular-arbuscular mycorrhizal fungi. Soil Sci. Soc. Am. J. 44:654-655.

41. Hussey, R. S., and Roncadori, R. W. 1978. Interaction of *Pratylenchus brachyurus* and *Gigaspora margarita* on cotton. J. Nematol. 10: 18-20.

42. Hussey, R. S., and Roncadori, R. W. 1982. Vesicular-arbuscular mycorrhizae may limit nematode activity and improve plant growth. Plant Dis. 66:9-14.

43. Ianson, D. C., and Linderman, R. G. 1991. Variation in VA mycorrhizal strain interactions with Rhizobium on pigeon pea. Pages 371-372 in: The Rhizosphere and Plant Growth, D. L. Keister and P. B. Cregan, eds., Kluwer Academic Publishers, Dordrecht, The Netherlands.

44. Ingham, R. E. 1988. Interactions betweeen nematodes and VA mycorrhizae. Agric., Ecosystems Environ. 24:169-182.

45. Jabaji-Hare, S. H., and Stobbs, L. W. 1984. Electron microscopic examination of tomato root coinfected with *Glomus* sp. and tobacco mosaic virus. Phytopathology 74:277-279.

46. Jalali, B. L., and Jalali, I. 1991. Mycorrhiza in plant disease control. Pages 131-154 in: Handbook of applied mycology. Soil and Plants. Vol. 1, D. K. Arora, B. Rai, K. G. Mukerji, and G. R. Knudsen, eds., Marcel Dekker, New York, NY.

47. Kaye, J. W., Pfleger, F. L., and Stewart, E. L. 1984. Interaction of *Glomus fasciculatum* and

Pythium ultimum on greenhouse grown poinsettia. Can. J. Bot. 62:1575-1579.

48. Krishna, K. R., Balakrishna, A. N., and Bagyaraj, D. J. 1982. Interactions between a vesicular-arbuscular mycorrhizal fungus and *Streptomyces cinnamomeous* and their effects on finger millet. New Phytol. 92: 401-405.

49. Krishna, K. R., Shetty, K. G., Dart, P. J., and Andrews, D. J. 1985. Genotype dependent variation in mycorrhizal colonization and response to inoculation of pearl millet. Plant Soil 86:113-125.

50. Linderman, R. G. 1986. Managing rhizosphere microorganisms in the production of horticultural crops. HortScience 21:1299-1302.

51. Linderman, R. G. 1988. Mycorrhizal interactions with the rhizosphere microflora: The mycorrhizosphere effect. Phytopathology 78:366-371.

52. Linderman, R. G. 1988. VA (Vesicular-Arbuscular) mycorrhizal symbiosis. ISI Atlas of Science, Animal and Plant Sciences Section 1:183-188.

53. Linderman, R. G. 1991. Mycorrhizal interactions in the rhizosphere. Pages 343-348 in: The Rhizosphere and Plant Growth, D. L. Keister and P. B. Cregan, eds., Kluwer Academic Publishers, Dordrecht, The Netherlands.

54. Linderman, R. G., Paulitz, T. C., Mosier, N. J., Griffiths, R. P., Loper, J. E., Caldwell, B. A., and Henkels, M. E. 1991. Evaluation of the effects of biocontrol agents on mycorrhizal fungi. Page 379 in: The Rhizosphere and Plant Growth, D. L. Keister and P. B. Cregan, eds., Kluwer Academic Publishers, Dordrecht, The Netherlands.

55. MacGuidwin, A. E., Bird, G. W., and Safir, G. R. 1985. Influence of *Glomus fasciculatum* on *Meloidogyne hapla* infecting *Allium cepa*. J. Nematol. 17: 389-395.

56. Meyer, J. R., and Linderman, R. G. 1986. Response of subterranean clover to dual inoculation with vesicular-arbuscular mycorrhizal fungi and a plant growth-promoting bacterium,

Pseuodmonas putida. Soil Biol. Biochem. 18: 185-190.

57. Meyer, J. R., and Linderman. R. G. 1986. Selective influence on populations of rhizosphere or rhizoplane bacteria and actinomycetes by mycorrhizas formed by *Glomus fasciculatum.* Soil Biol. Biochem. 18: 191-196.

58. Morandi, D., Bailey, J. A., and Gianinazzi-Pearson, V. 1984. Isoflavonoid accumulation in soybean roots infected with vesicular-arbuscular mycorrhizal fungi. Physiol. Plant Pathol. 24:357-364.

59. Morris, P. F., and Ward, E. W. B. 1992. Chemoattraction of zoospores of the soybean pathogen, *Phytophthora sojae*, by isoflavones. Physiol. Mol. Plant Pathol. 40:17-22.

60. Nelson, C. E. 1987. The water relations of vesicular-arbuscular mycorrhizal systems. Pages 71-91 in: Ecophysiology of VA Mycorrhizal Plants. G. R. Safir, ed., CRC Press, Inc., Boca Raton, FL.

61. Oliveira, A. A. R., and Zambolim, L. 1987. Interacao entro o fungo endomicorrizico *Glomus etunicatum* e o nematoide de galhas *Meloidogyne javanica* em feijoeiro com raiz partida. Fitopatol. Bras. 12: 222-225.

62. Papavizas, G. C., ed., 1981. Biological control in crop production. Allanheld, Osmun Publishers, Totowa, NJ.

63. Paulitz, T. C., and Linderman, R. G. 1989. Interactions between fluorescent pseudomonads and VA mycorrhizal fungi. New Phytol. 113:37-45.

64. Paulitz, T. C., and Linderman, R. G. 1991. Lack of antagonism between the biocontrol agent *Gliocladium virens* and vesicular arbuscular mycorrhizal fungi. New Phytol. 117:303-308.

65. Pond, E. C., and Menge, J. A. 1984. Improved growth of tomato in salinized soil by vesicular-arbuscular mycorrhizal fungi collected from saline soils. Mycologia 76:74-84.

66. Poss, J. A., Pond, E., Menge, J. A., and Jarrell, W. M. 1985. Effect of salinity on mycorrhizal onion and tomato in soil with and without

additional phosphate. Plant Soil 88:307-319.
67. Rambelli, A. 1973. The rhizosphere of mycorrhizae. Pages 299-343 in: Ectomycorrhizae. G. L. Marks and T. T. Kozlowski, eds., Academic Press, New York, NY.
68. Rosendahl, S. 1985. Interactions between the vesicular-arbuscular mycorrhizal fungus *Glomus fasciculatum* and *Aphanomyces euteiches* root rot of peas. Phytopath. Z. 114: 31-40.
69. Saleh, H., and Sikora, R. A. 1984. Relationship between *Glomus fasciculatum* root colonization of cotton and its effect on *Meloidogyne incognita*. Nematologica 30: 230-237.
70. Schenck, N. C. 1983. Can mycorrhizae control root diseases? Plant Dis. 65:230-234.
71. Schenck, N. C. 1987. Vesicular-arbuscular mycorrhizal fungi and the control of fungal root diseases. Pages 179-191 in: Innovative Approaches to Plant Disease Control, I. Chet, ed., John Wiley & Sons, Inc., New York, NY.
72. Schenck, N. C., and Kellam, M. K. 1978. The influence of vesicular arbuscular mycorrhizae on disease development. Fla. Agric. Exp. Stn. Tech. Bull. 798, Gainesville, FL.
73. Schonbeck, F. 1979. Endomycorrhiza in relation to plant diseases. Pages 271-280 in: Soil-borne plant pathogens. B. Schippers and W. Gams, eds., Academic Press, New York, NY.
74. Secilia, J., and Bagyaraj, D. J. 1987. Bacteria and actinomycetes associated with pot cultures of vesicular-arbuscular mycorrhizas. Can. J. Microbiol. 33: 1069-1073.
75. Sitaramaiah, K., and Sikora, R. A. 1982. Effect of the mycorrhizal fungus *Glomus fasciculatus* on the host-parasite relationship of *Rotylenchulus reniformis* in tomato. Nematologica 28: 412-419.
76. Smith, G. S. 1988. The role of phosphorus nutrition in interactions of vesicular-arbuscular mycorrhizal fungi with soilborne nematodes and fungi. Phytopathology 78:371-374.
77. Stewart, E. L., and Pfleger, F. L. 1977. Development of poinsettia as influenced by endomycorrhizae, fertilizer and root rot

pathogens *Pythium ultimum* and *Rhizoctonia solani*. Florist's Rev. 159:37, 79-80.

78. Strobel, N. E., Hussey, R. S., and Roncadori, R. W. 1982. Interactions of vesicular-arbuscular mycorrhizal fungi, *Meloidogyne incognita*, and soil fertility on peach. Phytopathology 72:690-694.

79. Suresh, C. K., Bagyaraj, D. J., and Reddy, D. D. R. 1985. Effect of vesicular-arbuscular mycorrhiza on survival, penetration and development of root-knot nematode in tomato. Plant Soil 87:305-308.

80. Suslow, T. V., and Schroth, M. N. 1982. Role of deleterious rhizobacteria as minor pathogens in reducing crop growth. Phytopathology 72:111-115.

81. Tisdall, J. M. 1991. Fungal hyphae and structural stability of soil. Aust. J. Soil Res. 29:729-743.

82. Vancura, V., Orozco, M. O., Grauova, O., and Prikryl, Z. 1989. Properties of bacteria in the hyphosphere of a vesicular-arbuscular mycorrhizal fungus. Agric., Ecosystems Environ. 29:421-427.

83. Wick, R. L., and Moore, L. D. 1984. Histology of mycorrhizal and nonmycorrhizal *Ilex crenata* 'Helleri' challenged by *Thielaviopsis basicola*. Can. J. Plant Pathol. 6: 146-150.

84. Zambolim, L., and Schenck, N. C. 1983. Reduction of the effects of pathogenic, root-infecting fungi on soybean by the mycorrhizal fungus, *Glomus mosseae*. Phytopathology 73: 1402-1405.

85. Zambolim, L., and Schenck, N. C. 1984. Effect of *Macrophomina*, *Rhizoctonia*, *Fusarium* and the mycorrhizal fungus *Glomus mosseae* on nodulated and non-nodulated soybeans. Fitopatologia Brasileira 9: 129-138.

ROLE OF ECTOMYCORRHIZAL FUNGI IN BIOCONTROL

Luc C. Duchesne
Forestry Canada, Chalk River,
Ontario, Canada

Ectomycorrhizal fungi are known to enhance the uptake of water and plant nutrients, increase resistance to root pathogens, and promote plant growth (22,24,30,46,49). Initial evidence for the role of ectomycorrhizal fungi in disease suppression was provided by a number of field observations that showed seedlings or trees of both angiosperms and gymnosperms to be more resistant to root pathogens than their nonmycorrhizal counterparts (34). Subsequently, the potential of ectomycorrhizal fungi as biological control agents was investigated, with emphasis on the control of damping-off in forest nurseries.

As with other biological control methods of soil-borne diseases (64), the major problem associated with the use of ectomycorrhizal fungi as biocontrol agents is one of predictability. To date, the protective influence of ectomycorrhizal fungi has been found to be extremely variable, therefore, unreliable for large-scale biocontrol. It is thought that a clearer understanding of ectomycorrhizal fungal physiology will help optimize the use of these organisms as biocontrol agents (5,15).

The objective of this paper is twofold. First, a summary of the major data pertaining to disease protection by ectomycorrhizal fungi as biological control agents is presented. Second, future research steps necessary to facilitate the large-scale application of ectomycorrhizal fungi as biological control agents are discussed.

MECHANISMS OF DISEASE SUPPRESSION BY ECTOMYCORRHIZAL FUNGI

Disease protection by ectomycorrhizal fungi has been associated with several active mechanisms that may either act singly, simultaneously, or synergistically (22,34,53,67). The mechanism of protection by ectomycorrhizal fungi is associated with: a) antibiosis; b) the synthesis of fungistatic/antifungal substances by roots in response to ectomycorrhizal fungi in a manner analogous to phytoalexin accumulation; and c) a physical barrier effect caused by the fungal mantle around ectomycorrhizal roots. These modes of action show inter- and intra-specific variations. In addition, Zak (64) postulated that a superior immobilization of nutrients by ectomycorrhizal fungi in the rhizosphere may be associated with disease protection. This hypothesis has yet to be proven. It is also important to note that disease protection by ectomycorrhizal fungi was observed in the absence of ectomycorrhizal roots (5, 13,14,16,19,61). For example, the ectomycorrhizal fungus *Paxillus involutus* (Batsch. ex. Fr) Fr. can protect seedlings of red pine (*Pinus resinosa Ait.)* against Fusarium root rot as early as two days following ectomycorrhizal inoculation, long before there is physical contact between roots and fungal hyphae (16,18).

The mechanisms listed above affect pathogens through direct activity: inhibition, competition, and mechanical or chemical exclusion. However, it is possible that other mechanisms, this time passive, are also involved in diseases suppression. Passive means of disease suppression include increased host vigor and modification of the basic physiology of plant roots after they encounter ectomycorrhizal fungi. The role of tree health in resistance to pathogens has been discussed at length (32), and it is possible that the promotion of water and nutrient uptake by ectomycorrhizal fungi affects the course of pathogenesis. Ectomycorrhizal fungi are known to synthesize plant growth regulators such as auxins and ethylene (22,39,50) that may alter the susceptibility of tree roots to pathogens.

FACTORS INFLUENCING DISEASE SUPPRESSION
BY ECTOMYCORRHIZAL FUNGI

Disease protection by ectomycorrhizal fungi is affected by environmental factors, tree species, pathogen species, inoculation timing, and the isolates of ectomycorrhizal fungi (4,13,42,43,45,51,52).

Influence of Environmental Factors
Biological control of diseases of crop plants is affected by soil fertility (8). It is speculated that disease suppression by ectomycorrhizal fungi is also altered by edaphic conditions (15,42,43). Three lines of evidence support this hypothesis. First, because pathogenesis by root pathogens is altered by host vigor and age, temperature, and nutrient and light levels (2,7,17,56,57,58,63,66), one should expect factors that favor pathogenesis or reduce host health to decrease the efficacy of ectomycorrhizal inocula. Second, environmental factors such as soil nutrient status, pH and root exudates are known to directly influence the metabolism of ectomycorrhizal fungi (3, 14,55). Marx et al. (36) observed that nitrogen and phosphorus fertilization of loblolly pine (*Pinus taeda* L.) seedlings decreases colonization by the ectomycorrhizal fungus *Pisolithus tinctorius* (Pers.) Coker & Couch. Ectomycorrhizal formation was favored by high sucrose content in feeder roots. It was then concluded that high root sucrose content is influenced by soil fertility which, in turn, influences ectomycorrhiza formation (36). Also, root exudates of red pine stimulated antibiosis by *P. involutus* resulting in protection against Fusarium root rot (11). In addition, high and low soil temperatures are known to alter ectomycorrhizae formation (40,55).

Consequently, proper manipulation of field conditions should favor disease protection by ectomycorrhizal fungi. For this, however, soils have to be managed under conditions that optimize both disease suppression and tree development. Such a hypothesis supports the necessity of a better

understanding of events occurring in the
mycorrhizosphere (31) to make more reliable the use of
ectomycorrhizal fungi in the field. Emphasis should
be given to understanding the precise effect of all
environmental parameters, particularly soil moisture
content and nutrient status, on disease suppression.
For example, two lines of evidence suggest that
enhanced antibiosis by *P. involutus* may be achieved
through soil amendments. First, it was demonstrated
that the synthesis of oxalic acid by *P. involutus* is
influenced by the bicarbonate, nitrogen, and calcium
contents of the growth medium (29). Oxalic acid
production by *P. involutus* was suppressed by ammonium
but stimulated by nitrate (29). Second, it was
observed that root exudates of red pine seedlings
stimulate the synthesis of antifungal compounds,
including oxalic acid, by *P. involutus* (11).

Influence of the Tree Genome

There are two general conclusions regarding the
influence of tree genetics on the effectiveness of
ectomycorrhizal fungi as biocontrol agents. A first
group of observations shows that the disease
suppressive effect of ectomycorrhizal fungi is more or
less uniform for all tree species. Marx and Davey
(35) observed suppression of *Phytophthora cinnamomi*
Rand. root rot of loblolly pine and short leaf pine
(*Pinus echinata* Mill.) using the ectomycorrhizal fungi
Leucopaxillus cerealis var. *piceina*, (Peck) ined. *P.
tinctorius*, and *Suillus luteus* (Fr.) S.F. Gray. There
was little variation in the level of protection
provided by the ectomycorrhizal fungi on the two pine
species. Sampangi et al. (52) found little or no
difference in the protection of seedlings of Norway
spruce (*Picea abies* L.) and Douglas fir (*Pseudotsuga
menziesii* (Mirb.) Franco) inoculated with the
ectomycorrhizal fungi *Laccaria bicolor* (Maire) Orton
and *Laccaria laccata* (Scop. ex Fr.) Cook.

Another group of studies demonstrated
host-generated inter-and intraspecific variations in
the effectiveness of ectomycorrhizal inocula. Perrin
and Garbaye (44) reported a reduction in the growth of
Pythium ultimum Thow in the rhizosphere of *Fagus* sp.

seedlings but not on *Quercus* sp. seedlings inoculated
with the ectomycorrhizal fungus *Hebeloma*
crustuliniforme (Bull. exst. Amans) Quel. Ross and
Marx (48) inoculated two races of sand pine (*Pinus
clausa* (Chapm. ex Engelm.) Vasey ex Sarg) with *P.
tinctorius* and *P. cinnamomi*. Protection against the
pathogen was observed on mycorrhizal seedlings of one
of the two races of sand pine (48). Inoculation of
seedlings of Jack pine (*Pinus banksiana* Lamb.),
eastern white pine (*Pinus strobus* L.), scotch pine
(*Pinus sylvestris* L.) and red pine with *P. involutus*
resulted in the protection of red pine and to a lesser
extent protection of scotch pine but not the
protection of Jack pine and Eastern white pine against
Fusarium root rot (13). Specificity in the
suppressive influence of *P. involutus* was ascribed to
differences in the stimulatory effects of pine
exudates (13).

     In the future, it will be critical to further
understand the influence of tree genomes in disease
protection by ectomycorrhizal fungi. In the
particular example of *P. involutus*, it is important to
isolate the stimulatory factor(s) present in red pine
root exudates. One application of such knowledge is
the modification of the edaphic environment in order
to stimulate antibiosis by *P. involutus* in the absence
of red pine root exudates. In ectomycorrhizal
associations where the mechanisms of disease
protection are different, other means of enhancing
protection may be devised.

Influence of Inoculation Timing and Inoculum Level

     The timing of inoculation with an ectomycorrhizal
fungus is critical for disease suppression.
Biological control was observed when inoculation of
seedlings with an ectomycorrhizal fungus was performed
either simultaneously (6) or prior to [as long as 11
weeks (57)] infection with a pathogen. There is no
instance reported where ectomycorrhizal fungi
protected seedlings after the onset of pathogenesis.
The general principle governing the timing of pathogen
infection following ectomycorrhizal inoculation is
that the ectomycorrhizae within the host must reach

the physiological stage at which disease suppression is expressed before the onset of pathogenesis. It has been shown (L. Duchesne, unpublished results) that the protective effect of *P. involutus* is doubled in red pine when infection with the pathogen is carried out one to four days following inoculation, as compared to seedlings inoculated simultaneously with the pathogen and ectomycorrhizal fungus. In red pine, root infection by *Fusarium oxysporum* Schlecht. begins as early as two days following inoculation (18), and the suppressive influence of *P. involutus* is also first detected at the same time (16). Ectomycorrhizal fungi may start protecting a host faster if they do not need root formation to initiate the protective mechanism (see Mechanisms of Disease Suppression in this paper).

It is likely that the inoculum level of ectomycorrhizal fungi necessary for disease control will vary according to the disease conducive potential of soils which, in turn, is influenced by the overall environmental conditions, pathogen virulence, and the number of pathogen propagules. Unfortunately, these factors have not been studied at the research or the operational level.

Influence of the Pathogen Genome

Control of seedling root rot caused by *F. oxysporum* and *F. moniliforme* Sheld. was observed on red pine inoculated with *P. involutus* (5,12), but the latter was ineffective against *Cylindrocarpon destructans (Zinssmeister) Scholten* (5). Strobel and Sinclair (59) observed protection of Douglas fir seedlings by *L. bicolor* against a nonaggressive isolate of *F. oxysporum*, whereas there was no protective effect against two aggressive isolates of this pathogen. Comparison of the effect of antibiosis by ectomycorrhizal fungi on several strains of root pathogenic fungi shows inter- and intraspecific variation (34,47). The practical implication of these observations is that most ectomycorrhizal fungi show a limited action range, making the selection of single isolates for widespread use difficult.

Influence of the Ectomycorrhizal Isolate

The inoculation of seedlings of Norway spruce and Douglas-fir with three isolates of *L. laccata* showed differences in the order of 250% in the protection level provided by the best isolate (51,52). Comparison of the protective influence of four isolates of *P. involutus* on Fusarium root rot of red pine showed a protective effect for one isolate but none for the remaining three isolates (L. Duchesne, unpublished results). In vitro analyses of antibiosis by ectomycorrhizal fungi show variations within isolates of the same fungal species (47). Comparison of the synthesis of fungitoxic substances in the F1 progeny of *P. tinctorius* shows a large degree of variation among different sibling isolates (25). Moreover, circumstantial evidence suggests substantial intraspecific variation in disease suppression mechanisms because at least two different isolates of *P. tinctorius* exhibit different mechanisms of disease suppression (6,7,35).

These observations are consistent with the finding that ectomycorrhizal fungi show various levels of inter- and intraspecific variation in aggressiveness (65). Wong and Fortin (65) reviewed several stages of root colonization where genetic control is necessary for the onset and maintenance of the ectomycorrhizal symbiosis. Molina and Trappe (38) grouped ectomycorrhizal fungi into three specificity classes. Fungi such as *L. laccata, P. tinctorius*, and *P. involutus* show little host specificity because they can colonize a large number of plant species; fungi such as *Alpova diplophloeus* (Zeller & Dodge) Trappe and *Rhizopogon vinicolor* Smith show specificity to one host genus; and a third class includes those fungi with intermediate degrees of host specificity. However, one should expect that the host range for optimal suppressive influence and the host range for enhanced growth are not necessarily identical; it is likely that a number of different groups of genes are involved in these two events. It is important to characterize the degree of genetic disparity between these two processes to optimize the beneficial effects of ectomycorrhizal fungi.

FUTURE RESEARCH AND DEVELOPMENT EMPHASES

There are four avenues of research and development that may facilitate the large-scale application of ectomycorrhizal fungi as biocontrol agents: a) the selection of ectomycorrhizal fungi with wide action ranges; b) use of ectomycorrhizal fungi in conjunction with other means of disease protection; c) the use of natural pesticides derived from ectomycorrhizal fungi; and d) genetic engineering of ectomycorrhizal fungi for enhanced disease suppression.

Selection of Ectomycorrhizal Fungi with Wide Action Ranges

In theory it is possible to select ectomycorrhizal fungus isolates suitable for field use against a wide variety of pathogens, on numerous tree species, under various field conditions because, as discussed earlier, some ectomycorrhizal fungi show a wide host range (38). Selection should be aimed at isolates that display as many of the following features as possible (15): a) mass-production should be readily and inexpensively achieved for long periods of time without loss of biocontrol potential; b) the organism's biology should be amenable to storage and transportation prior to field application; c) field application should require little manipulation; d) selected isolates should display their protective effect soon after inoculation; e) they should compete efficiently with the indigenous soil microflora and colonize soils extensively under variable field conditions; f) they must suppress many pathogenic species; and g) they should not disrupt normal beneficial soil processes (15). Ectomycorrhizal fungus isolates possessing all of these attributes have not yet been found and, so, in practice the selection of ectomycorrhizal isolates for biocontrol should entail some degree of compromise between effectiveness and operational disadvantages.

In addition to the seven features discussed above, the selection of suitable ectomycorrhizal fungi as biocontrol agents is made even more problematic because of our lack of knowledge on how the many biological and environmental variables influence disease suppression both in the laboratory and the field. This is common to many biocontrol agents (20). A thorough review of the selection methods for the field use of ectomycorrhizal fungi was presented in Duchesne et al. (15). Ideally, a screening program should be conducted under all operational field conditions, but this is difficult to achieve. Consequently, an alternative three-stage screening procedure has been proposed (15). Such screening should be conducted first in the laboratory on a large number of isolates, followed by a greenhouse investigation of selected fungal isolates, and, finally, in the field using only the most promising isolates. Despite numerous shortcomings, this procedure should allow an accurate assessment of the potential of a large number of isolates with relatively modest resources.

Ectomycorrhizal Fungi and Other Means of Disease Protection

On a commercial scale, it may be possible to increase the spectrum of action of ectomycorrhizal fungi by using mixed inocula or groups of organisms with different ranges of action (15,31). This may prove an efficient way of providing versatile inocula with extended field efficacy; pathogen populations are expected to adapt faster to pesticides with single modes of actions than to pesticides with multiple modes of action.

Biological control agents of plant diseases do not yet reach efficiencies comparable to that of chemical pesticides for which resistance has not been acquired (5). Therefore, dual control systems using both chemical and biological control agents was proposed by Chakravarty et al. (5). The combined use of *P. involutus* and the fungicides benomyl and oxine oxalate showed a synergistic effect that increased suppression of root rot caused by *F. oxysporum* (5). This kind of integrated disease management can be performed only if the environmental impact of the chemical pesticide is minimal and if the ectomycorrhizal fungal inocula are tolerant to the chemicals.

Use of Pesticides Derived from Ectomycorrhizal Fungi

It may be possible to use antibacterial/ antifungal substances synthesized by ectomycorrhizal fungi for controlling root diseases in forest nurseries (15,21,26). More than 100 species of ectomycorrhizal fungi are known to synthesize such substances (10,22,34). These chemicals are termed naturally-occurring pesticides (natural pesticides) and constitute an alternative to synthetics (41). One example is provided by the use of oxalic acid for controlling Fusarium root rot of red pine (14). Oxalic acid was isolated as a toxic substance synthesized by *P. involutus* and tied to its suppressive influence against Fusarium root rot (14). Treatment of red pine seedlings with authentic oxalic acid protects them against Fusarium root rot in vitro (14). Although the efficiency of oxalic acid as a pesticide in the field and its effects on ecosystems have yet to be determined, this result illustrates that research on antibiosis by ectomycorrhizal fungi may lead to the discovery of new pesticides. Recently, others have reported attempts to isolate and develop natural pesticides synthesized by ectomycorrhizal fungi (26,60). A similar approach is being used to develop natural pesticides from plant extracts (37) and it is possible that phytoalexin-like compounds synthesized by tree roots in response to ectomycorrhizal fungi will also be investigated as potential natural pesticides.

A number of conditions must be met for a natural pesticide to be effective at the operational level: a) it should have no effect on other environmental components than its target pathogens; b) it should be non-phytotoxic at dosages that are effective for disease control; c) it should be stored and shipped without needs for special facilities; d) it should be efficient against a wide spectrum of root pathogens; and e) its structure should be amenable to inexpensive chemical synthesis. Evidently, the impact of such substances on the environment must be investigated as thoroughly as in the case of chemical pesticides.

Genetic Engineering of Ectomycorrhizal Fungi

Genetic engineering of ectomycorrhizal fungi offers the potential to increase the efficiency of these microorganisms as biological control agents. Genomic transformation of *L. laccata* was achieved using a marker gene encoding for resistance to the drug hygromycin B (1). These results are promising but there is a need to extend further the gene transfer technology to other ectomycorrhizal fungi and to isolate target genes that should be altered in the fungal genome. Understandably, the genetic improvement of ectomycorrhizal fungi is only possible if these traits and/or their regulatory mechanisms are encoded by single or a limited number of loci. Aspects of the ectomycorrhizal symbiosis that can be targeted for genetic improvement are aggressiveness, specificity, growth, antibiosis, and competitivity in soils. Compatibility in ectomycorrhizal associations is genetically controlled both in plants (65) and in ectomycorrhizal fungi (27,28). Protein profiles of *P. tinctorius* and *Eucalyptus globulus* Labill. show that the regulation of hundreds of peptides is altered by the onset of the ectomycorrhizal symbiosis (23). In vivo 35S-L-methionine protein labelling of *P. involutus* and the root of red pine shows that suppression of Fusarium root rot by this ectomycorrhizal fungus is associated with differential protein synthesis (11). Moreover, the same experiment showed differential protein expression by both *P. involutus* and red pine roots as early as two days following inoculation, and this without contact between the two organisms.

CONCLUSIONS

In the past, the protective effect of ectomycorrhizal fungi was investigated mainly against root-rot pathogens of conifer seedlings in forest nurseries. Conifer stocks constitute a major group for reforestation and are easier to manipulate in vitro than most hardwood species. Evidence has yet to be presented for a direct protective effect of ectomycorrhizal fungi against other organisms than root pathogens.

One problem associated with the use of ectomycorrhizal fungi in forest nurseries is that fungal inocula will have to remain in soils for the duration of seedling stay in a particular nursery bed. There is a danger, under these circumstances, that soil microorganisms, pathogen and non-pathogen alike, can acquire resistance to the protection factors of ectomycorrhizal fungi. Duchesne et al. (15) postulated that knowledge of the factors that influence disease protection may also be advantageous because proper field manipulation may avoid exposing subepidemic populations of pests and potential pests to control mechanisms. This may lengthen the useful life of biological control agents.

At present, it is not possible to predict the usefulness of ectomycorrhizal fungi in disease control at the operational level. However, this is a relatively young field of research when compared to the biological control of diseases on crops. Despite the difficulties associated with ectomycorrhizal fungi, their large scale use shows potential both as a component of integrated pest management programs and for their environmental friendliness.

LITERATURE CITED

1. Barret, V., Dixon, R. K., and Lemke, P. A. 1990. Genetic transformation of a mycorrhizal fungus. Appl. Microbiol. Biotechnol. 33:313-316.
2. Bloomberg, W. J. 1973. Fusarium root rot of Douglas-fir. Phytopathology 63:337-341.
3. Brundrett, M. 1991. Mycorrhizas in natural ecosystems. Adv. Ecol. Res. 21:271-315.
4. Chakravarty, P., and Hwang, S. F. 1991. Effect of an ectomycorrhizal fungus, *Laccaria laccata*, on Fusarium damping-off in *Pinus banksiana* seedlings. Eur. J. For. Pathol. 21:97-106.
5. Chakravarty, C., Peterson, R. L., and Ellis, B. E. 1991. Interaction between the ectomycorrhizal fungus *Paxillus involutus*, damping-off fungi and *Pinus resinosa*. J. Phytopathol. (Berl.) 132:207-218.

6. Chakravarty, P., and Unestam, T. 1986. Role of mycorrhizal fungi in protecting damping-off of *Pinus sylvestris* seedlings. Pages 811-814 in: Proc. 1st Eur. Symp. Mycorrhizae, 1985. V. Gianinazzi-Pearson and S. Gianinazzi, eds., Dijon, France.

7. Chakravarty, P., and Unestam, T. 1987. Differential influence of ectomycorrhizae on plant growth and disease resistance in *Pinus sylvestris* seedlings. J. Phytopathol. (Berl.) 120:104-120.

8. Chen, W., Hoitink, H. A., Schmitthenner, A. F., and Tuovinen, O. H. 1988. The role of microbial activity in suppression of damping-off caused by *Pythium ultimum*. Phytopathology 78:314-322.

9. Duchesne, L. C. 1989. Protein synthesis in *Pinus resinosa* and the ectomycorrhizal fungus *Paxillus involutus* prior to ectomycorrhiza formation. Trees 3:73-77.

10. Duchesne, L. C., Peterson, R. L., and Ellis, B. E. 1987. The accumulation of plant-produced antimicrobial compounds by ectomycorrhizal roots: a review. Phytoprotection 68:17-47.

11. Duchesne, L. C., Peterson, R. L., and Ellis, B. E. 1988a. Pine root exudate stimulates antibiotic synthesis by the ectomycorrhizal fungus *Paxillus involutus*. New Phytol. 106:558-262.

12. Duchesne, L. C., Peterson, R. L., and Ellis, B. E. 1988b. Interaction between the ectomycorrhizal fungus *Paxillus involutus* and *Pinus resinosa* induces resistance to *Fusarium oxysporum*. Can. J. Bot. 66:558-562.

13. Duchesne, L. C., Campbell, S. E., Koehler, H., R.L. Peterson, R. L., and Ellis, B. E. 1989a. Pine species influence suppression of Fusarium root rot by the ectomycorrhizal fungus *Paxillus involutus*. Symbiosis 7:139-148.

14. Duchesne, L. C., Ellis, B. E., and Peterson, R. L. 1989b. Disease suppression by the ectomycorrhizal fungus *Paxillus involutus*: contribution of oxalic acid. Can. J. Bot. 67:2726-2730.

15. Duchesne, L. C., Peterson, R. L., and Ellis, B. E. 1989c. The future of ectomycorrhizal fungi as biological control agents. Phytoprotection

70:51-58.

16. Duchesne, L. C., Peterson, R. L., and Ellis, B. E. 1989d. The time course of disease suppression and antibiosis by the ectomycorrhizal fungus *Paxillus involutus*. New Phytol. 111:693-398.

17. Elad, Y., Chet, I., and Katan, J. 1980. *Trichoderma harzianum*: a biocontrol agent effective against *Sclerotium rolfsii* and *Rhizoctonia solani*. Phytopathology 70:119-121.

18. Farquhar, M. L., and Peterson, R. L. 1990. Early effects of the ectomycorrhizal fungus *Paxillus involutus* on the root rot organism Fusarium associated with *Pinus resinosa*. Can. J. Bot. 68:1589-1596.

19. Farquhar, M. L., and Peterson, R. L. 1991. Later events in suppression of Fusarium root rot of red pine seedlings by the ectomycorrhizal fungus *Paxillus involutus*. Can. J. Bot. 69:1372-1383.

20. Fravel, D. R. 1988. Role of antibiosis in the biocontrol of plant diseases. Annu. Rev. Phytopathol. 26:75-91.

21. Garrido, N., Becerra, J., Marticorena, C., Oehrens, E., Silva, M., and Horak, E. 1982. Antibiotic properties of ectomycorrhizal fungi and saprophytic fungi growing on *Pinus radiata* D. Don. Mycopathologia 77:93-98.

22. Harley, J. L., and Smith, S. E. 1983. Mycorrhizal Symbiosis. Academic Press, New York. 483 pp.

23. Hilbert, J. L., and Martin, F. 1988. Regulation of gene expression in ectomycorrhizas. I. Protein changes and the presence of ectomycorrhiza-specific polypeptides in the *Pisolithus-Eucalyptus* symbiosis. New Phytol. 110:339-346.

24. Jeffries, P. 1987. Use of mycorrhizae in agriculture. CRC Crit. Rev. Biotechnol. 5:319-357.

25. Kope, H. H., and Fortin, J. A. 1991. Genetic variation in antifungal activity by sibling isolates of the ectomycorrhizal fungus *Pisolithus arhizus*. Soil Biol. Biochem. 23:1047-1051.

26. Kope, H. H., Tsantrizos, S., Fortin, J. A., and Ogilvie, K. K. 1991. p-Hydroxybenzoylformic acid

and (R)-(-)-p-hydroxymandelic acid, two antifungal compounds isolated from the liquid culture of the ectomycorrhizal fungus *Pisolithus arhizus*. Can. J. Microbiol. 37:258-264.

27. Kropp, B., and Fortin, J. A. 1988. The incompatibility system and ectomycorrhizal performance of monocaryons and reconstituted dikaryons of *Laccaria bicolor*. Can. J. Bot. 66:289-294.

28. Kropp, B., McAfee, B. J., and Fortin, J. A. 1987. Variable loss of ectomycorrhizal ability in monokaryotic and dikaryotic cultures of *Laccaria bicolor*. Can. J. Bot. 65:500-504.

29. Lapeyrie, F., Chilvers, G. A., and Bhem, C. A. 1987. Oxalic acid synthesis by the mycorrhizal fungus *Paxillus involutus* (Batsch. ex. Fr) Fr. New Phytol. 106:139-146.

30. Le Tacon, F., Garbaye, J., and Carr, G. 1987. The use of mycorrhizas in temperate and tropical forests. Symbiosis 3:179-206.

31. Linderman, R. G. 1987. Perspectives on ectomycorrhiza research in the Northwest. Pages 72-74 in: Mycorrhizae in the Next Decade: Practical Applications and Research Priorities. Proc. 7th NACOM, 1987. D. M. Sylvia et al., eds., Gainesville, FL.

32. Manion, P. D. 1981. Tree Disease Concepts. Prentice Hall, NJ.

33. Marx, D. H. 1969. The influence of ectotrophic ectomycorrhizal fungi on the resistance to pathogenic infections. I. Antagonism of mycorrhizal fungi to pathogenic fungi and soil bacteria. Phytopathology 59:153-163.

34. Marx, D. H. 1973. Mycorrhizae and feeder root diseases. Pages 351-382 in: Ectomycorrhizae: Their Ecology and Physiology. G. C. Marks and T. T. Kozlowski, eds., Academic Press, New York.

35. Marx, D. H., and Davey, L. B. 1969. The influence of ectotrophic mycorrhizal fungi on the resistance of pine roots to pathogenic infections. III. Resistance of aseptically formed mycorrhizae to infection by *Phytophthora cinnamomi*. Phytopathology 59:549-558.

36. Marx, D. M., Hatch, A. B., and Mendicino, J. F. 1977. High soil fertility decreases sucrose content and susceptibility of loblolly pine roots to ectomycorrhizal infection by *Pisolithus tinctorius*. Can. J. Bot. 55:1569-1574.
37. Nakatani, N., Yamada, Y., and Fuwa, H. 1987. 7-geranyloxycoumarin from juice oil of *Citrus hassabu* and antimicrobial effects of related coumarins. Agric. Biol. Chem. 51:419-427.
38. Molina, R., and Trappe, J. M. 1982. Patterns of ectomycorrhizal host specificity and potential among Pacific Northwest conifers and fungi. For. Sci. 28:423-458.
39. Nylund, J.A. 1988. The regulation of mycorrhiza formation ---carbohydrate and hormone theories reviewed. Scand. J. For. Res. 3:465-479.
40. Perrido, H. Oliva, M., and Huber A. 1983. Environmental factors determining the distribution of *Suillus luteus* fructifications in *Pinus radiata* grazing-forest plantations. Plant Soil 71:367-370.
41. Perkins, J. H. 1985. Naturally occurring pesticides and the pesticide crisis. Pages 297-328 in: Handbook of Natural Pesticides: Methods. N. B. Mandava, ed., CRC Press, Boca Raton, FL.
42. Perrin, R. 1985. Peut-on compter sur les mycorhizes pour lutter contre les maladies des plantes ligneuses? Eur. J. For. Pathol. 15:372-379.
43. Perrin, R. 1985. L'aptitude des mycorhizes ê protÄger les plantes contre les maladies: panacÄe ou chimere? Ann. Sci. For. (Paris) 42:453-470.
44. Perrin, R., and Garbaye, J. 1983. Influence of ectomycorrhizae on infectivity of Pythium-infected soils and substrates. Plant Soil 71:345-351.
45. Perrin, R., and Nouveau, M. 1985. L'association mycorrhizienne de Pinus sylvestris avec l'*Hebeloma crustuliniforme* et *Laccaria laccata* et les maladies causees par le Pythium spp. Pages 793-798 in: Proc. 1st Eur. Symp. Mycorrhizae, V. Gianinazzi-Pearson and S. Gianinazzi, eds., Dijon, France.

46. Peterson, R. L., Piche, Y., and Plenchette, C. 1984. Mycorrhizae and their potential use in the agricultural and forestry industries. Biotechnol. Adv. 2:101-120.

47. Pratt, B. 1971. Isolation of basidiomycetes from eucalypt forests and assessment of their antagonism to *Phytophthora cinnamomi*. Trans. Br. Mycol. Soc. 56:243-250.

48. Ross, E. W., and Marx, D. M. 1972. Susceptibility of sand pine to *Phytophthora cinnamomi*. Phytopathology 62:1197-1200.

49. Ruehle, J. L., and Marx, D. H. 1979. Fiber, food, fuel and fungal symbionts. Science (Wash. D.C.) 206:419-422.

50. Rupp, L. A., Mudge, K. W., and Negm, F. B. 1989. Involvement of ethylene in ectomycorrhiza formation and dichotomous branching of roots of mugo pine seedlings. Can. J. Bot. 67:477-482.

51. Sampangi, R., and Perrin, R. 1985. Attempts to elucidate the mechanisms involved in the protective effect of *Laccaria laccata* against *Fusarium oxysporum*. Pages 807-810 in: Proc. 1st Eur. Symp. Mycorrhizae, V. Gianinazzi-Pearson and S. Gianinazzi, eds., Dijon, France.

52. Sampangi, R., Perrin, R., and Le Tacon, F. 1985. Disease suppression and growth promotion of Norway spruce and Douglas-fir seedlings by the ectomycorrhizal fungus *Laccaria laccata* in forest nurseries. Pages 799-806 in: Proc. 1st Eur. Symp. Mycorrhizae, V. Gianinazzi-Pearson and S. Gianinazzi, eds., Dijon, France.

53. Schisler, D. A., and Linderman, R. G. 1989. Selective influence of volatiles purged from coniferous forest and nursery soils on microbes of a nursery soil. Soil Biol. Biochem. 21:389-396.

54. Sinclair, W. A., Cowles, D. P., and Hee, S. M. 1975. Fusarium root rot of Douglas-fir seedlings: suppression by soil fumigation, fertility management, and inoculation with spores of the fungal symbiont *Laccaria laccata*. For. Sci. 21:390-399.

55. Slankis, V. 1974. Soil factors influencing formation of mycorrhizae. Annu. Rev. Phytopathol.

12:437-457.
56. Smiley, R. W., and Cook, R. J. 1973. Relationship between take-all of wheat and rhizosphere pH in soils fertilized with ammonium- vs nitrate-nitrogen. Phytopathology 63:882-890.
57. Stack, R. W., and Sinclair, W. A. 1975. Protection of Douglas-fir seedlings against Fusarium root rot by a mycorrhizal fungus in absence of mycorrhiza formation. Phytopathology 65:468-472.
58. Strobel, N. E., and Sinclair, W. A. 1991. Influence of temperature and pathogen aggressiveness on biological control of Fusarium root rot by *Laccaria bicolor* in Douglas-fir. Phytopathology 81:415-420.
59. Strobel, N. E., and Sinclair, W. A. 1991. Role of flavanolic wall infusions in the resistance induced by *Laccaria bicolor* to *Fusarium oxysporum* in primary roots of Douglas-fir. Phytopathology 81:420-425.
60. Suh, H. W., Crawford, D. L., Korus, R. A., and Shetty, K. 1991. Production of antifungal metabolites by the ectomycorrhizal fungus *Pisolithus tinctorius* strain SMF. J. Ind. Microbiol. 8:29-36.
61. Sylvia, D. M., and Sinclair, W. A. 1983. Suppressive influence of *Laccaria laccata* on *Fusarium oxysporum* and on Douglas-fir seedlings. Phytopathology 73:384-389.
62. Sylvia, D. M., and Sinclair, W. A. 1983. Phenolic compounds and resistance to fungal pathogens induced in primary roots of Douglas-fir seedlings by the ectomycorrhizal fungus *Laccaria laccata*. Phytopathology 73:390-397.
63. Weste, G., and Marks, G. C. 1987. The biology of *Phytophthora cinnamomi* in Australasian forests. Annu. Rev. Phytopathol. 25:207-229.
64. Whipps, J. M. 1986. Use of micro-organisms for biological control of vegetable diseases. Aspects Appl. Biol. 12:75-94.
65. Wong, K. K. Y., and Fortin, J. A. 1990. Root colonization and intraspecific mycobiont variation in ectomycorrhiza. Symbiosis 8:197-231.

66. Wright, E., Harvey, G., and Bigelow, C. 1963.
    Tests to control Fusarium root rot of ponderosa
    pine in the Pacific Northwest.   Tree Planters'
    Notes 14:15-20.
67. Zak, B. 1964.   Role of mycorrhizae in root
    disease.   Annu. Rev. Phytopathol. 2:377-392.

ROLE OF VAM FUNGI IN MINE LAND REVEGETATION

F.L. Pfleger, E.L. Stewart, and R.K. Noyd
University of Minnesota
St. Paul, Minnesota

Mining for natural resources generates a variety of wastes that differ in their biological, chemical, and physical attributes. Factors such as soil fertility, pH, moisture, temperature and soil microbial activity influence successful plant establishment, growth, and survival on such wastes (77). Mining activities have a deleterious effect on plant and soil microbial community health that must be ameliorated with the addition of amendments in order to effect successful revegetation. Amendments provide a nutrient source and an environment that enhances the development of a fully structured and functional soil microbial community (76). Functional soil microbial communities are composed of a complex of species that differ in their environmental tolerances, physical requirements and habitat adaptations (59). The soil microbial community functions in a number of critical biogeochemical processes that are fundamental to soil development and plant growth. The objectives of mine waste reclamation include re-establishment of a sustainable plant community that is functional and aesthetically pleasing. These restored landscapes directly benefit by serving as wildlife habitats, water and wind erosion abatement, rejuvenation of aquifers, and where applicable provide for agricultural and forestry production.

A major component of the soil microbial community is a group of fungi that form a symbiotic association with the roots of most terrestrial plants known as vesicular-arbuscular mycorrhizae (VAM). These fungi are among the most ubiquitous soil organisms found in terrestrial ecosystems (18). They form symbioses with a broad range of plant species and can contribute to plant growth and survival by reducing stresses associated with nutrition, water/aeration, soil structure, pH, salt, toxic metals and biotic factors such as pathogens, hyphal feeders, and organic matter (74). Although studies involving the occurrence of VAM fungi on mining wastes are numerous, few studies have attempted to understand the interaction between VAM fungi and plant communities in revegetation success. Most current operational revegetation practices ignore VAM-plant community dynamics and have had limited success in establishing self-sustaining plant communities. There is an urgent need to understand VAM-plant community interactions on disturbed and undisturbed systems and to apply this knowledge in developing more innovative, cost effective and successful revegetation practices.

The objective of this chapter is to establish a basis in the literature for (a) how plant succession and diversity are linked to VAM re-establishment in disturbed habitats, (b) evaluation of crucial aspects that must be considered when selecting VAM fungi for revegetation, (c) criteria for plant selection for revegetation, and (d) approaching revegetation strategies associated with specific types of mining activities and associated spoil. The spoil types included in this chapter are coal, oil sand, oil shale, iron ore, and rock phosphate.

PLANT SUCCESSION, VAM RE-ESTABLISHMENT, AND DIVERSITY

Early successional weedy annuals that colonize mining wastes are often non-mycorrhizal. Initial attempts at revegetation with mycotrophic species may be unsuccessful because of low levels of VAM fungal inoculum found in disturbed soils (49). Reeves et al. (62) and Janos (32) hypothesized that, under these conditions secondary succession will likely be slowed, resulting in prolonged longevity of early successional non-mycorrhizal plants. Allen (4) has suggested that the rate of plant succession on a disturbed site may be dependent on the rate at which VAM fungal inoculum increases with time. Additionally, Allen et al. (10) showed that an inoculum mix of *Glomus fasciculatum* (Thaxter) Gerd. & Trappe emend. Walker & Koske, *Glomus mosseae* (Nicol. & Gerd.) Gerd. & Trappe, *Glomus microcarpum* Tul. & Tul., and *Gigaspora margarita* Becker & Hall invaded the roots of a nonmycotrophic, early successional weed Russian thistle (*Salsola kali* L.), reducing plant growth and survival. Some nonmycotrophic weed species have been shown to facilitate early succession of mycotrophic plants through shading, capture of moisture, and wind protection (6). If VAM reduces the vigor of the facilitator species, early succession is likely to be prolonged. However, a reduction in density of Russian thistle caused by VAM would, in turn, potentially reduce competition for nutrients and water (5) thereby allowing for seral succession of mycotrophic species (8).

Observations on natural plant succession in severely disturbed sites can provide important information for improving reclamation success. Miller (49) noted that western wheat grass (*Agropyron smithii* Rydb.) was nonmycorrhizal in a severely disturbed site in Wyoming, while in most habitats, grasses form VAM. Janos (32) has stated that many grasses are facultative mycotrophs and may not be solely dependent upon mycorrhizal endophytes. Allen (3) and Allen (7) observed that mature western wheat grass grew as a facultative mycotroph in severely disturbed sites. An awareness that some grasses can be facultative mycotrophs, with low mycorrhizal dependency, is helpful when selecting early successional plant species for revegetation programs (45).

Daft and Nicolson (22) found that all the grasses and most dicotyledonous plants growing on coal tips in England were colonized by VAM. Species richness however, was low in all sites. Mycorrhizal formation and plant diversity on mine spoils has been shown to increase as plant succession occurs. Allen and Allen (4) showed that percent root colonization in revegetated mine sites increased to approximately one-half as much as an undisturbed grassland prairie within two to three years. Stockpiling topsoil and a dilution effect caused by incorporation of surface materials from a depth of as much as 2 m could account for the low colonization they observed. Schwab and Reeves (66) reported that in an undisturbed sage community, very little VAM fungal inoculum occurred below a soil depth of 0.5 m. Loree and Williams (46) observed that mean root density, degree of colonization, and spore frequency in 5-7 year-old sites in south eastern Wyoming were comparable to those of an adjacent undisturbed site. When Lambert and Cole (40) used topsoil from a grassy field incorporated into mine spoil, high levels of colonization occurred on white clover (*Trifolium repens* L.) after a single growing season. Site variability, distribution of VAM inoculum, and hosts might account for the marked differences observed by Allen and Allen (4), Lambert and Cole (40), and Loree and Williams (46). Similarly, Mott and Zuberer (54) seeded coastal bermudagrass (*Cynodon dactylon* (L.) Pers.) into mixed overburden replacement soil and found that percentage root colonization had returned to premining levels after 3 to 7 years. They also noted a 10-fold increase in spore numbers in unmined vs. mined soils in the same time period.

It is not clear what factors influence the rate of recolonization of coal mining wastes by VAM fungi. However, factors such as severity of disturbance, harshness of the site, low inoculum levels (49), edaphic characteristics, plant species, cover and time since reclamation (4) certainly influence the rate of VAM fungal recolonization. It is clear that VAM fungal inoculum levels are inherently important in the establishment and maintenance of a healthy plant community.

ASPECTS OF VAM SELECTION FOR REVEGETATION

Ecological specialization of VAM fungi has been demonstrated to occur at various taxonomic levels. This fact highlights the importance of selecting VAM fungi that may be ecologically adapted to a particular disturbed site for use in revegetation programs. This specialization has been demonstrated to be in response

to several factors including soil pH, presence of
heavy metals, source of inoculum, and host endophyte
specificity.

Edaphic Factors

Lambert and Cole (40) used a mixture of isolates
of VAM fungi selected from revegetated low pH spoils
to establish growth rate studies at a low and high pH
level with white clover. Biomass production was
greatest at the low pH, suggesting that ecological
adaptation had occurred. Porter et al. (60) studied
Acaulospora laevis Gerd. & Trappe and a *Glomus* spp.
originally isolated from low and high pH soils,
respectively. When reciprocal inoculations were done
with these species in high and low pH soils they
colonized and sporulated only in the soils of origin.
They suggested that pH regulated spore germination and
hyphal out growth. Additional evidence of adaptation
was provided by Morton and Walker (52) who described
Glomus diaphanum Morton & Walker associated with
acidic coal mine spoils high in aluminum. Studies
such as these beg the question regarding plant and
fungal adaptation and the use of agronomic rather than
locally adapted native plants in conjunction with
ecologically efficient VAM fungi. The evidence that
these fungi can differentially affect plant growth
demonstrates that intraspecific ecological
specialization exists (15,47).

Kiernan et al. (37) studied coal spoils that had
naturally revegetated over a twenty year period and
found the composition of VAM fungal species to be
highly diverse. Koske (39) also found that VAM fungal
species composition was higher in stabilized (older)
sand dunes than in mobile (younger) dunes. Johnson et
al. (35) found that the VAM fungal species composition
and richness was correlated to edaphic variation and
plant productivity in a chronosequence of 1 to 60
years in Minnesota. Their research showed that the
fungal community shifted during secondary succession
of abandoned fields based on a field to forest
chronosequence. However, the total number of VAM
fungal species did not change over time, but richness
did. These studies clearly show that natural

revegetation of disturbed sites can occur over time
with high species richness and diversity of VAM fungi.
They also indicate that to accelerate revegetation of
mine spoils it is necessary to use inputs that
facilitate plant and microbial succession. For
example, Lambert and Cole (40) suggested that the
introduction of VAM fungi would be beneficial if
indigenous fungi are absent or unadapted, available P
is low, and plants used in revegetation positively
respond to VAM.

Heavy Metals

Continuous exposure of plant populations to soils
contaminated with heavy metals can eventually result
in the selection of tolerant plants (12,61). Likewise,
the selection of VAM fungi tolerant to heavy metals
was found by Gildon and Tinker (30) who reported that
roots of leek (*Allium porrum* L.) plants growing on a
site heavily contaminated with zinc and cadmium had
high levels of mycorrhizal colonization. Furthermore,
they showed that colonization by *Glomus mosseae*
associated with heavy metal contaminated soil was not
affected when exposed to various levels of zinc and
cadmium, whereas colonization by a *G. mosseae* isolate
from non-contaminated soil was greatly reduced or
nearly eliminated at higher levels of these heavy
metals. Inoculations with the heavy metal-tolerant
mycorrhizal isolate did not result in higher
accumulation of these metals in plant tissue even
though the soil had high levels of metals.

Killham and Firestone (38) reported that heavy
metal concentration was higher and plant biomass
production lower in bunchgrass (*Ehrharta calycina*
Thanb.) colonized by *Glomus fasciculatum* than in
non-mycorrhizal plants. Moreover, they noted that as
soil acidity increased, there was a corresponding
increase in heavy metal uptake in VAM plants. Best et
al. (14) noted that several plant species completely
lacked VAM when growing in acid soil containing high
levels of Al, even in the presence of abundant
propagules, but were heavily mycorrhizal in adjacent
non-acid soils. Some of the plant species growing in
the acid areas were mycorrhizal, but appeared much

less healthy growing there than in non-acid areas.
The unhealthiness of plants may be attributed to the
increased uptake of Al by VAM fungi. The species of
VAM fungi, their intraspecific variation, cultural
conditions and edaphic factors under which the fungus
and plants were growing could account for the
different experimental results obtained by these
workers. These results show that some plants and VAM
fungi develop tolerance to heavy metals. Combinations
of these plant and VAM fungal species should be
selected when revegetating spoils contaminated with
heavy metals.

Inoculum Source and Density

Indigenous VAM fungi are likely candidates for
reinoculation in reclamation efforts. Kahn (36),
Lindsey et al. (44), and Stahl et al. (69) used
indigenous VAM fungi as inoculants to study the
effects on plant growth on mine spoils. Kahn (36)
used *Glomus macrocarpum* Tul. & Tul., *G. mosseae*, and
Sclerocystis rubriformis Gerd.& Trappe originally
isolated from coal tips, and the E3 strain of *Endogone*
obtained from Rothamsted Experimental Station to
compare the effects on growth of onions in coal wastes
under greenhouse conditions. In this study, the
nonindigenous E strain fungus was reported to be the
most efficient among all species examined, while *G.
mosseae* was the most efficient among the indigenous
species, and nearly equal to the E strain fungus.
Similarly, Lindsey et al. (44) collected *G.
fasciculatum* and *G. mosseae* from soil around rabbit
brush (*Chrysothamnus nauseosus* (Pall.) Britton) and
fourwing saltbush (*Atriplex canescens* (Pursh) Nutt.)
in an undisturbed site adjacent to a spoil and used
these fungal species to inoculate the aforementioned
hosts grown in the greenhouse. These fungi had a
positive effect only on the growth of rabbit brush. At
the end of the study, *G. fasciculatum* was the most
dominant fungus found in the rhizosphere of rabbit
brush roots, whereas neither VAM fungal species became
established in fourwing saltbush (44). They suggested
that the lack of VAM colonization may be due to host
endophyte specificity. Interestingly, in a different

study, Aldon (1) reported that *G. mosseae* was able to colonize roots, increase survival, and enhance growth of fourwing saltbush on spoil banks.

Stahl et al. (69) collected mycorrhizal fungi from sagebrush (*Artemisia tridentata* subsp. *wyomingensis* Beetle & Young) from undisturbed sites adjacent to a spoil and from a mixed overburden topsoil that had been stockpiled for seven years. *Entrophospora infrequens* (Hall) Ames & Schneider, *G. fasciculatum*, *G. macrocarpum*, *G. microcarpum*, and *G. mosseae* were found in the undisturbed site, whereas only *G. fasciculatum* and *G. microcarpum* were found in the disturbed, reclaimed soil. Spore levels in the undisturbed and disturbed sites were 19 and 0.5 propagules per gram of soil, respectively. The VAM fungal species from the undisturbed sites were used to inoculate sagebrush directly seeded on site in the field, as well as plants grown in containers and then transplanted to the field site. Mycorrhizal colonization in the directly seeded treatments decreased from 63 to 22% in 15 months, while colonization in the noninoculated controls increased from 8 to 22% over the same time period. Colonization in the transplanted container grown plants decreased from 82 to 19%, and increased from 4 to 21% in the noninoculated controls. They noted that after nearly two years there were no significant differences in seedling growth, survival, and colonization levels among inoculated and control treatments. It is noteworthy that 2.5 years after reclamation, three of the five introduced VAM fungal species were not found in the disturbed soil. Two species, *G. fasciculatum* and *G. microcarpum*, were common to both sites. It is plausible that the two taxa remaining were those that occurred originally in the disturbed soil at extremely low levels. If this is the situation, these two taxa demonstrate a degree of site and host adaptation. Moreover, this may also explain the colonization equilibrium that had developed after a period of 15 months (Table 1).

Many species of VAM fungi occur over a broad range of habitats and environmental conditions. What this strongly suggests is that these fungi are capable

Table 1.   Effect of VAM fungi on infection, establishment, survival and plant height on *Artemisia tridentate* subsp. *wyomingensis* seeded or transplanted into disturbed reclaimed soil.

	Seeded		Container-grown*		Seeded		Container-grown	
	+Inoc.	-Inoc.	+Inoc.	-Inoc.	+Inoc.	-Inoc.	+Inoc.	-Inoc.
	Spring 1984				Autumn 1984			
Mycorrhizal infection level† (%)	63±12a‡	8±5b	8±10a	4±2b	25±12	19±10	38±15	17±8
Establishment§ (%)	1	1	---	---	---	---	89	94
Survival∥ (%)	100	100	100	100	97	98	89	94
Plant Height (cm)	6·8±4·1	7·3±4·5	11·8±5·0a	6·9±3·2b	7·2±4·2	7·6±4·5	12·2±6·3	9·9±5·5
	Spring 1985				Autumn 1985			
Mycorrhizal infection level (%)	21±13	17±10	20±9	18±10	22±9	22±10	19±11	21±9
Survival (%)	93	96	88	92	93	95	87	92
Plant height (cm)	7·8±4·4	7·9±5·0	12·7±6·7	11·8±5·8	8·2±4·8	8·5±5·5	13·1±7·0	12·3±6·4

*   Observations of container-grown plants in spring, 1984 were made prior to transplantation into the field site.
†   Infection level is expressed as the mean percentage of the total examined root length infected by VA mycorrhizal fungi.
‡   For infection level and plant height data, values given are the treatment mean ± standard deviation. Entries followed by different letters are statistically different at the 95% level of confidence as determined by two samples t tests. Only data within seeded or container-grown treatments at the same sampling date were compared.
§   Establishment data are given for the seeded treatment as the percentage of live sagebrush seed planted resulting in seedlings by the spring of 1984. For the container-grown treatment they are given as the percentage of planted seedlings surviving the first growing season in the field.
∥   Survival data are given as the percentage of initially established plants remaining alive at each sampling date.

From P. D. Stahl, S.E. Williams and Martha Christensen, New Phytol. (1988), 347-354.

of genetic differentiation at the intraspecific level (23,68). It is not surprising that intraspecific variation would occur in VAM fungi since examples of this phenomenon for other nonVAM fungi have been noted (17,42).

Waaland and Allen (80) examined how the rate of plant succession on surface mined sites in Wyoming was affected by VAM fungal spore densities. They looked at the relationship between plant cover, mycorrhizal colonization levels, and spore densities in several sites representing different age classes and land use including: undisturbed, reclaimed soil, 1-6 year old spoils, and 10 to 31 year old orphan soils all without top soil applications. They found no

correlation between site age and spore densities, or
between percentage root infection and time. They did
observe greater spore densities and higher percentage
root infection in undisturbed versus disturbed soils.
In this study, spore densities never reached levels
similar to those found in the undisturbed sites.
Consequently, Waaland and Allen (80) suggest that
practices such as topsoiling with stored material and
direct hauling may be more important than time since
disturbance in successful revegetation attempts.

Because of low VAM fungal inoculum densities
associated with mine spoils, Allen (9) and Allen and
MacMahon (11) suggest that a possible alternative to
inoculating vast tracts of land would be to establish
islands of mycorrhizal plants from which
recolonization of the reclaimed spoil might occur more
rapidly. Plants in the islands may also serve as
catchments for VAM fungal spores that are wind blown
onto a disturbed site as well as providing protective
habitats for animals and insects (83) that can vector
spores and mycelial fragments.

PLANT SELECTION FOR REVEGETATION

Plant types selected for revegetation play an
important role in modifying soil development. Miller
and Jastrow (51) have developed a conceptual model
that depicts the role of plant roots and mycorrhizal
fungal hyphae in promoting soil aggregation. Rothwell
and Eagelston (64) have shown that selection of plant
species such as C-4 plants can modify soil development
on surface mines through improved soil structure and
the formation of soil aggregates. These plants have a
high growth rate and release high levels of
carbohydrates into the rhizosphere that stimulates
microbial activity, including VAM fungi. Activity of
soil microbes and VAM fosters stabilization of soil
aggregates (63) that function to reduce soil erosion
and help recreate a nutrient reserve (50).

Lambert and Cole (40) showed the importance in
revegetation of using legume plants. Mycorrhizal
flatpea (*Lathyrus sylvestris* L.), birdsfoot trefoil
(*Lotus americanus* (Nutt.) Bisch.), and crown vetch

(*Coromilla varia* L.) showed a highly significant increase in growth over nonVAM plants, and survival was 6 to 15 fold greater than for nonVAM plants after two winters. Furthermore, they showed that plant biomass production in the presence of minimal P and VAM was greater than with nonVAM plants grown at higher P levels.

In the southwestern United States, recently stored topsoil contains a seed bank that includes fourwing saltbush as a prevalent species. The viability of this seed decreases with time in stored topsoil. Therefore, spreading the stored topsoil as soon as possible results in the early establishment of this plant type on mine waste spoils. As a facultative mycotroph (3,7), this plant can establish and proliferate in the absence of VAM fungi, yet become mycorrhizal in the presence of VAM fungi. Ultimately this plant species aids in the increase of the mycorrhizal inoculum potential (MIP) levels as shown by Aldon (2).

The use of nursery propagated VAM plants is one method of introducing suitable plant fungal combinations into mine wastes. This method will most likely be effective if the nursery plants are propagated in the presence of ecologically-adapted VAM fungi which are known to have a positive effect on plant survival and growth on reclaimed sites. Roots of several different woody plant species propagated in nurseries for outplanting in surface mines have been found to be colonized by VAM fungi. Colonization levels were high, even though high rates of fertilizer were used and the nursery soils were fumigated with methyl bromide. Root samples taken after one year from nursery plants transplanted into a mine waste, revealed reduced levels of colonization compared to levels observed on nursery grown plants (13). An example of this approach to revegetation has been demonstrated by Davies and Call (26). They inoculated Chinese tallow tree (*Sapium sebiferum* (L.) Roxb.) and *Sophora secundiflora* plants native to the area, with a mixture of VAM fungal species obtained from mixed overburden followed by outplanting into a mine spoil. Growth and development was enhanced, but improved survival was noted only for *S. secundiflora* after two years. Non-inoculated plants of the same species transplanted on the same site showed very low levels of colonization after two years.

SPECIFIC MINING ACTIVITIES AND RECLAMATION

Coal

Coal mining spoils are poor habitats for plant establishment, growth and survival. They characteristically have low nutrient and organic matter content, low water holding capacity, and low pH (78). In addition, erosion and extremes in surface temperature are also impediments to revegetation. Rothwell and Eagleston (64) addressed the importance and benefits of microbial activity to successful revegetation on mine spoil. They suggested that in situations where top soil is shallow, highly eroded, or leached, replacement of that material with a topsoil substitute that includes an array of beneficial microorganisms may be required.

The addition of organic residues (such as mineral overburden) and mineral fertilizer to mining wastes usually results in improved plant growth and survival and increases the potential for successful revegetation of disturbed sites. This may occur directly and indirectly through increased fertility by biological decomposition, stabilization of pH, increasing the cation-exchange capacity, and binding of metal ions (20). The physical properties of spoils are also affected by the addition of organic matter by improving soil aggregation and structure, water holding capacity, and modification of soil temperature. Addition of amendments to mining wastes have a role in re-establishment of heterotrophic soil. Clearly, the use of organic residues on selected spoil is a critical component in the reconstitution of a functional plant and soil microbial community (71,77).

Mine spoils devoid of topsoil have adverse chemical and physical properties that can persist for many years. Such sites are not considered biologically stable, and as such, revegetation attempts are difficult. Mine spoils that received topsoil had similar chemical, physical and biological properties to that of an adjacent undisturbed site within 3 to 5 years. In a study by Fresquez et al. (28), disturbed non-topsoiled and undisturbed sites had comparable populations of bacteria, ammonium oxidizers and *Streptomyces*, but differed in diversity of fungal genera within the spoil after 8 years. They suggested that the low diversity of fungi found in non-topsoiled spoil may result in reduced enzymatic activity that reduces normal nutrient cycling and decomposition, thus slowing the process of soil microbial community development to its pre-disturbance composition (29) (Table 2). In addition to VAM fungi, these studies demonstrate the importance of re-introducing all components of the soil microbial community back into the soil during revegetation attempts on mine spoils. We believe there is a need to examine the tandem role of VAM and other microbes in re-establishing vegetation on mine spoils. Therefore, the addition of amendments as an initial carbon and nutrient source is a means of quickly

TABLE 2.   Distribution and relative density of fungal genera of an undisturbed soil and reclaimed coal mine spoils and soils[1].

Fungal groups	Undis-turbed	Non-topsoiled			Topsoiled				
		1974	1975	1976	1978	1979	1980	1981	1982
Acremonium	1	0	0	0	0	14	2	0	0
Alternaria	1	0	0	1	0	0	1	6	8
Annellophorella	1	0	0	0	0	0	0	0	0
Aspergillus	5	26	68	83	37	14	30	16	10
Chaetomium	2	1	4	0	6	18	39	1	1
Chrysosporium	0	0	2	0	0	1	1	0	0
Cladosporium	0	1	0	0	0	0	0	0	0
Curvularia	15	0	5	4	3	2	13	6	2
Dreshlera	1	0	0	0	0	0	0	0	0
Fusarium	7	2	8	11	4	15	21	26	13
Humicola	5	0	2	0	5	0	0	0	0
Myrothecium	1	0	0	0	1	0	38	17	14
Mortierlla	1	0	0	0	0	3	0	0	0
Mucor	0	0	1	2	0	2	0	2	1
Mycelia sterilia	0	3	13	1	1	0	1	0	0
Penicillium	15	51	25	38	7	7	5	39	69
Phoma	5	0	0	0	1	0	0	0	0
Rhizopus	5	0	1	0	1	2	0	5	2
Sepedomium	0	0	0	0	0	1	1	0	0
Stachybotrys	3	1	0	0	4	8	11	25	18
Stemphyllium	0	0	0	0	1	0	4	0	0
Thielavia	1	0	0	0	0	0	0	0	0
Trichoderma	3	0	0	0	0	0	0	0	0
Trichurus	0	0	0	0	0	0	0	0	1
Unidentified									
Isolate No. 1	15	0	0	0	0	0	0	1	0
Isolate No. 2	3	0	0	0	0	2	1	0	0
No. of isolates	90	85	129	140	71	89	168	144	139
No. of genera	19	7	10	7	12	13	14	11	11
Fungal diversity	1.091	0.448	0.650	0.476	0.748	0.946	0.876	0.857	0.715
Evenness	0.853	0.530	0.650	0.563	0.693	0.849	0.764	0.823	0.687

[1]Relative density of isolates from six 1:1000 soil dilution plates.

From: P.R. Fresquez and E. F. Aldon and W.C. Lindemann, Reclamation and Revegetation Research, 4(1986)245-258.

re-establishing the soil microflora and microfauna (65).

Lindemann et al. (43) noted that if mine spoils in the arid West are not topsoiled prior to reclamation attempts, the addition of an amendment will be necessary to establish and stimulate soil microbial activity. They examined the effect of spoil amendments on soil microbial parameters and VAM fungal colonization of grasses. Their research demonstrated that the addition of either sewage sludge or hay or topsoiling at a depth of 30 cm increased the number of soil microorganisms, enzyme activity and fungal genera distribution in nonrhizosphere spoil. Of the amendments used, covering the spoil with 30 cm of topsoil provided the greatest VAM colonization.

Spoil plots amended with either mineral fertilizer, peat or liquid sewage sludge and seeded with slender wheat grass (*Agropyron trachycaulum* (Link) Malte.) were established to examine the effects on VAM development (84). Evidence of VAM formation was found after 2 weeks in peat-amended plots, whereas colonization was not detected in plots amended with fertilizer or sewage sludge until 6 and 10 weeks after planting, respectively. Colonization levels remained highest in the peat-amended spoil after four growing seasons (85). These researchers speculated that heavy metals in the plots amended with sewage sludge delayed VAM formation. They found that *Glomus aggregatum* (Schenck & Smith) emend. Koske and *G. mosseae* were the most common VAM fungi associated with slender wheat grass. None of the amendments had an effect on *G. mosseae* spore densities, whereas *G. aggregatum* spore density was greatest on peat and lowest in the sewage amended treatments. Spore size for *G. mosseae* found in the sewage-amended spoil was at the lower end of the range for the species. This observation warrants further experimentation to determine the significance of amendments and their effects on colonization and reproductive potential of VAM fungi. Moreover, the decision to use an organic amendment should be based on its composition and the effects it can have on the soil microfauna and microflora.

Oil Sand Spoils

Mining of surface oil sands involves the removal of vegetation and upper soil layers or overburden. The heavy crude oil is extracted with hot water and NaOH. This mining process generates massive amounts of extracted oil sand tailing, which are nearly biologically sterile, lacking essential plant nutrients, clay minerals, and organic matter. They are subject to wind and water erosion, have little water holding capacity, and are generally unstable. Revegetation strategies for these mining wastes include the routine use of amendments such as peat and mineral overburden.

The effects of shallow or deep overburden, peat, a mixture of overburden plus peat, and sand on the growth of slender wheat grass were evaluated under greenhouse conditions by Danielson et al. (27). They found that plant growth was best in the peat treatment layered 5 cm deep and poorest in the deeply layered overburden. This growth response was attributed to the greater availability of P to the roots in the shallow overburden, whereas the deeply layered overburden was both deficient in N and P and restricted root proliferation. The improved shoot growth of slender wheat grass observed with the application of a shallow overburden was attributed to the introduction of VAM fungal inoculum and an improved moisture holding capacity. In a subsequent study, Danielson et al. (24) found VAM fungal inoculum present in all four overburden types with colonization levels low in all but the shallow amended spoil. Even though VAM colonization was high in the shallow overburden amended spoils, none was observed in roots found in the underlying oil sand spoil. This finding is interesting, for what factors would prevent VAM hyphae from colonizing the root as they penetrate from the overburden into the underlying oil shale spoil? In addition to the VAM studies, soil microbial respiration was measured in each of the four overburden amendments. They (24) found that the level of microbial activity was related to the amount of organic matter present in the treatment. Peat had 79.1% organic matter and had the highest microbial activity and biomass. The tailing sand contained almost no organic matter, showed little microbial activity, and contained negligible microbial biomass. Mixing spoils with amendments tailored for a particular site may beneficially modify the rate and efficiency of the decomposition and mineralization of dead plant parts and enhance nutrient release from litter back to the primary producers. VAM fungi are a vital link between decomposers and nutrient acquisition by plants and are involved with the other members of the soil microbial community in maintaining ecosystem functioning.

Zak & Parkinson (85) looked at the effects of amendments on VAM development in roots of slender wheat grass growing on oil sands spoil at 2 and 4 years following amendment application. The amendments applied were peat, fertilizer, or sewage sludge. At 2 and 4 years, the peat-amended plots had the highest level of VAM, whereas low VAM was noted in plants growing in the fertilized oil sands. Plants growing in the sewage-amended plots did not become colonized during the first two growing seasons, but colonization was detected after the fourth year. They (85) attribute the low levels of VAM colonization to the high rates of extractable P detected in the sewage-treated spoil. Between the second and fourth year the VAM status of the plants in the oil sand spoil changed. For example, root length with VAM decreased in the peat-amended spoil, but still was highest among all treatments; VAM level in roots of plants on the fertilized plots increased, but remained less than the peat amended treatment. Most control plants had died by the end of the 4th year; among those plants that survived, root colonization had increased from 4 to 36%.

Oil Shale

The commercialization of and excavation of oil shale in the semi-arid regions of the western U.S. has created large amounts of processed shale which, like most mining wastes, is difficult to revegetate. The process of oil extraction from mined shale requires crushing and retorting at an extremely high temperature (500 C). The end product is dark colored, nutrient deficient, biologically sterile, lacking in structure, high in pH and highly saline (19).

Current methods used to revegetate oil shale wastes include the use of agronomic plants that require high inputs of fertilizer, as well as leaching of salts, topsoiling, mulching and irrigation. As noted with other mining disturbances, primary plant succession on oil shale spoil consists of nonmycorrhizal weedy annuals (19). This finding supports those of other workers studying revegetation on coal spoil (4,21,49,62). The invasion of these spoils by the nonmycotrophic species halogeton (*Halogeton glomeratus*, (M.) Bieb.), summer cyprus (*Kochia scoparia* (L.) Schrader), and Russian thistle is attributed in part to the extremely low number of VAM fungal propagules.

Bioassay of MIP in topsoil placed over deposited oil shale revealed colonization levels of approximately half those obtained in undisturbed soil (19). Call and McKell (20) transplanted container grown fourwing saltbush plants inoculated with indigenous VAM fungi into processed oil shale and disturbed native soil. The plants inoculated with VAM fungi had a greater above ground biomass, percent cover, and height than the noninoculated controls after two growing seasons. Phosphorous and water uptake was greater in the VAM plants than in the nonVAM controls. A significant reduction in VAM colonization from 13.0 to 3.8% occurred from the time of out-planting to harvest in the processed shale plantings. In spite of the low levels of colonization, a positive growth response was observed. Water-soluble constituents in the retorted oil shale likely contributed to the decline in VAM colonization in the inoculated plants growing in processed shale. These researchers also noted that transplanted nonVAM saltbush did not become mycorrhizal. In contrast, Zak & Parkinson (85) noted that noninoculated plants growing in fertilized oil sands were sporadically mycorrhizal after the second growing season, but were not different in growth from the non-fertilized controls. These studies underscore the importance of

VAM in the establishment of early and later successional plants in processed oil shale.

Studies by Reeves et al. (62) on shale wastes covered with subsoil and then topsoiled, resulted in higher MIP levels. They noted a relationship between the degree of disturbance and subsequent recovery of an ecosystem. For example, as the degree of disturbance increases, there is a corresponding increase in recovery time, in part due to reduced MIP levels. MIP levels were also shown to be influenced by fertilizer amendments for revegetation of oil shale deposits. High levels of nitrogen applied to disturbed topsoil resulted in decreased MIP, whereas low fertilizer levels had no detrimental effect on MIP and helped to insure successful plant establishment.

The effects of topsoil thickness, fertilizer, successional patterns of introduced and native plants, and soil biological activity of revegetated sites were studied over a 5 year period by Biondini et al. (16). They found that fertilizer and top soil thickness of 30 cm had a long lasting influence on plant species composition, and that plots established with these treatments were dominated by grasses whereas unfertilized plots of similar topsoil thickness were dominated by alfalfa (*Medicago sativa* L.).

Successful revegetation of oil shale deposits, as with other spoils generated by mining activities, will no doubt require an integrated approach such as the addition of fertilizer, organic amendments, use of stress tolerant plants, and VAM fungal species known to improve survival and growth of introduced plants.

Iron Ore

Iron ore is mined from large open pits and processed into marble-sized pellets called taconite and shipped to steel mills in that form. The mining of iron ore has resulted in large tracts of disturbed land that consist of two types of permanent waste materials, overburden and tailing. Taconite iron ore tailing are the non-ore-bearing portions of rock. Tailing are segregated into coarse and fine-sized particles and disposed onto the land's surface. Revegetation research has occurred on overburden sites

(70), coarse tailing (55-58,70) and fine tailing
(33,34,58). Tailing lack important soil properties
such as profile development and biological activity.
The coarse and fine tailing are very different media
for plant growth. Stresses associated with coarse
tailing include an alkaline pH of 7.5 to 8.3, a lack
of water-holding capacity, inadequate levels of
nitrogen and phosphorus, and a dark color that absorbs
heat. Fine tailing pose the same nutrient stresses,
but have a much greater water-holding capacity than
coarse tailing and thus support the establishment of
plant cover and cycling of nutrients.

On coarse iron tailing, nonmycotrophic ruderal
plant species cannot tolerate the stressful conditions
and do not persist. Late successional plant species
that are adapted to tolerate stressed habitats
generally are mycorrhizal and represent a possible
long term solution (62). There is evidence that
physiological responses in plant hosts can be related
to intraspecific variation of VAM fungi (15). This
view is supported by Stahl and Smith (68) who report
that different geographic isolates of *G. macrocarpum*
and *G. microcarpum* had different effects on the water
relations of the rangeland grass *A. smithii*. These
results have directed researchers to consider adjacent
undisturbed vegetation for locally-adapted sources of
VAM fungal inoculum. Noyd et al. (58) found *Glomus
intraradix* Schenck & Smith associated with plants in
undisturbed soils as well as in fine and coarse
tailing, suggesting that this species is adapted to a
wide range of habitats.

Allen (7) has shown that the rate of plant
succession on a disturbed site may depend on the rate
at which VAM fungal inoculum increases with time. Noyd
et al. (58) found that spore densities average 5.6
spores/gm and 1.0 spore/ gm on fine and coarse
tailing, respectively, 2 years after deposition.
They (58) studied the rate of VAM fungal spore
immigration into coarse tailing from adjacent
undisturbed vegetation and found rates averaging 2
spores/gm/month over a single growing season (Figure
1). These low levels may be one reason why Leisman
(41) found the rate of natural revegetation of coarse

iron spoil banks to be slow. On spoil banks as old as 31 years, he found only a sparse herbaceous ground cover.

On a smaller scale, there are instances where natural dispersal of VAM fungi is adequate to enhance the fitness of vegetation. Propagules of VAM fungi are dispersed by a variety of biotic agents (48,75,83). Plants colonized by VAM fungi were found growing in depressions in coarse tailing where leaves accumulate, and wolf scat containing VAM fungal spores was

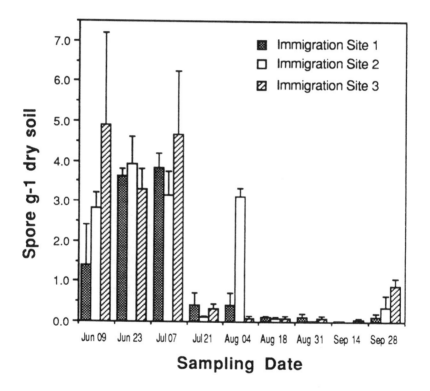

Figure 1. Immigration rate of VAM spores into coarse tailing USX Corporation in northern Minnesota. Data are pooled average values from 10 samples collected in 1990 from spore traps with an area of 0.25 m^2 (modified from Noyd et al. Amer. Soc. Surf. Min. Reclam. 1992, p. 361-369.

deposited (58). Large numbers of propagules were found on buckwheat (*Fagospyrum esculentum* Moench.) seed in a rodent seed cache (58,75). Islands of VAM plants, such as those occasionally observed on coarse tailing (58), may be excellent models for an island revegetation approach as advocated by Allen (9) and Allen and MacMahon (11). This strategy requires that suitably-adapted VAM fungal species shown to improve plant host biomass production and survival be selected. Unfortunately adoption of the island revegetation model is restricted by reclamation goals and regulations that measure success in terms of percent plant cover. For example, Minnesota requires a 90% plant cover within three years. Basing success on short term plant cover makes the development of long-term, practical cost-effective methodologies difficult.

Coarse iron tailing lack another important component of the below-ground ecosystem; organic matter. Tailoring an organic amendment to site-specific conditions is an important part of revegetation research and can influence ecosystem recovery (85). Ground cover has been improved by adding a variety of organic amendments such as sawdust (67), composted yard waste (56), and composted municipal solid waste (55-57). However, not all organic amendments have a beneficial effect on populations of VAM fungi. Johnson and McGraw (33) found that composted papermill sludge significantly increased plant biomass, spore densities, and levels of root colonization in less than two growing seasons. In contrast, tailing treated with straw mulch contained fewer spores. Some amendments, such as municipal solid wastes, contain very high levels of phosphorus. Adding such wastes to native soils would surely suppress VAM, but when incorporated into iron tailing the dilution factor results in acceptable P levels (56).

Phosphate Rock

Mining for rich supplies of phosphate involves removal of nonweathered overburden soil to depths ranging from 2-6 m (82). This creates a mosaic of

overburden mounds with extremely steep slopes
surrounded by water-filled canals (81). According to
Florida Administrative Code (Chapter 16C-16.0051,
amended Feb 2, 1987), 10% of phosphate mined lands are
to be returned to their natural state (72). Phosphate
mining has disturbed a wide variety of ecosystems that
range from mesic habitats dominated by oaks and scrub
to hydric areas dominated by cypress (82). Reclaiming
these mature forested areas to meet state code
requires an understanding of the ecology of the wide
assortment of vegetation and soils impacted by the
mining process.

Rock phosphate overburden has higher
concentrations of Ca, P, K, and Mg than premined
soils. Mined soils also contain considerably greater
proportions of clay to sand, and extremely low levels
of organic matter and nitrogen (82). Like other mine
spoils, overburden is generally considered to be
essentially abiotic. These harsh soil conditions
initially slow plant establishment, and in some cases
cause very persistent shrub communities to become so
well established that they prevent the invasion of
later successional forest species (81). This limits
the establishment of VAM plant species that form a
major proportion of late-successional plants. Since
revegetation success is measured by the degree to
which mine lands are restored to their natural state,
it is imperative that the role of VAM in the process
be fully understood. Adding and establishment of VAM
fungi in a revegetation strategy is a component of
natural ecosystem reconstruction (86).

Natural plant succession that had occurred on
smaller tracts of phosphate-mined lands afforded
Wallace and Best (81) an opportunity to investigate
below ground succession and VAM fungal dynamics in
different aged nonreclaimed areas that included sites
recently-mined to those 3, 8, 17, 43, and 60 years
old. They found that a majority of plants/roots at
the 3-year old site were colonized from 2 to 90% by
VAM fungi. These results indicates that VAM fungi can
invade the site allowing for VAM dependent species to
become established. Colonization at the 8-year old
site was generally the same as at the 3 year old site.

This data suggests that establishment of VAM is very heterogenous, possibly as the result of the difficulty in obtaining sufficient inoculum to form VAM.   As shrubs invaded and composed a greater proportion of the plants at the 17 year old site, they were found to be colonized by VAM fungi. At the 43-year old site, roots of vines, grasses, herbs and shrubs were heavily colonized.   Later successional vegetation at the 60-year old site was mostly colonized by ectomycorrhizal fungi.   Results indicated a reduction of VAM fungi as later plant succession resulted in more ectomycorrhizal hosts. Unfortunately, the identification of VAM fungal species as they occurred on the different aged sites was not included in this study.   Such information could add significantly to our understanding of VAM fungal succession. This type of information should be included in future ecological studies.

VAM fungi occurring on phosphate mined lands must tolerate and be adapted to high levels of P, which has been shown repeatedly to reduce VAM colonization (53). In soils of high P availability, how do VAM fungi benefit the host plant? It is reasonable to assume that VAM fungi also benefit host plants by increasing the absorption of other limiting nutrients and water. To determine the effect of *G. mosseae* and *Glomus occultum* Walker on trees from a variety of Florida ecosystems, Best et al. (14) directly seeded 25 tree species into microplots. Using 500 g of inoculum/plot, they found that for the first growing season these fungi had a positive effect on growth and seedling density.   The importance of VAM fungi did not carry over into the second growing season.   They attribute this decline to the dominance of pioneer tree species that are less dependent on VAM for growth and survival.

Sylvia (72) used three *Glomus* spp. from three different locations in a growth response study on swamp dogwood (*Cornus foemina* Mill.), gallberry (*Ilex glabra* L.), and sweet pepper bush (*Clethra alnifolia* L.). He found that VAM doubled shoot and root biomass of swamp dogwood.   Shoots of gallberry were five times larger than controls when inoculated with *G.*

intraradix; sweet pepper bush was not affected by VAM. He found no significant ecological or edaphic correlations between performance and the locations of the three sources. However, he (72) concluded that when woody plants are placed into soils low in VAM fungi, inoculation of mycorrhizal-dependent species such as swamp dogwood and gallberry should be considered. In another investigation, height and stem growth of swamp dogwood was nearly double that of controls when inoculated with *G. etunicatum*, *G. intraradix*, or a mixture of *Glomus* spp. during the container phase of plant production (73). However, well-colonized plants did not exhibit a significant increase in growth or survival 18 months after outplanting to a site that contained VAM fungi (73). This work helps reclamationists know the VAM dependencies of plants involved in revegetation and helps establish an inoculation program.

To establish an inoculation program, VAM fungi should be selected that are known to survive in post mined soils and to enhance plant biomass production. Wallace et al. (82) proposed similar concerns regarding locally occurring native plants that are expected to invade, survive, and successfully regenerate on newly created soil systems. Different VAM fungal species and intraspecific variants can influence the physiology of the host plant differently (15). Wallace et al. (82) assessed differences of growth enhancement of sweetgum (*Liquidambar styraciflua* L.) by *G. macrocarpum* and a composite sample of several VAM fungal species isolated from phosphate mine clay. The composite sample was assumed to be adapted to edaphic factors of the mine soil. Results indicated that the composite sample promoted greater growth than either the control or *G. macrocarpum* treatments at all nutrient levels. Leaf areas of plants treated with the composite were 30 times greater than controls and 1.7 times greater than those inoculated with *G. macrocarpum*. These results are significant in illustrating the efficient combinations of native plants and adapted sources of VAM for use in revegetation.

CONCLUSIONS

The revegetation of a variety of mine land disturbances is dependent upon understanding how soil components, amendments, and the soil microbial community, including VAM fungi, influence plant establishment and succession. In greenhouse experiments, VAM fungi have been shown to confer many benefits that enhance plant growth. Despite many attempts, however, demonstrations of increased survival and growth of plants with and without VAM under field conditions are few. This indicates that there is insufficient information regarding how to successfully manage VAM fungi in disturbed habitats. Furthermore, it points out the need to allocate more resources to support VAM/revegetation field research. We think this is important, because later successional plants adapted to stresses that limit plant growth, form VAM and are likely to benefit in these ecologically stressed environments. Understanding how plants chosen for revegetation interact with VAM is important. Plants exhibit a wide range of dependency on these fungi and different geographical isolates can have profound differences on growth and survival of each plant species. Information concerning the status of VAM dependency on native plants used in revegetation is limited and needs further study. Once this information is available, the challenge is to find plant/VAM combinations that enhance revegetation efforts.

Mining wastes differ depending on the parent geologic material and the processes used for extraction. Likewise, climatic conditions associated with mining activities differs from one geographic region to another and can greatly influence the methods used for revegetation. Additional research is needed in this area, and the results incorporated into revegetation strategies specific to these geographical regions.

Mining activities result in decreased populations of VAM fungal propagules (33). Stored topsoils should be managed to minimize the loss of VAM propagules. This can be accomplished by direct hauling, avoiding

deep stock piled soils, and minimizing storage time.
These practices enhance survival of VAM fungi and
maintain a more fully structured soil microbial
community. Both factors are important in soil
aggregate formation leading to improved soil
structure, nutrient capture and re-establishment of
vegetation on severely disturbed sites.

The ability to produce large quantities of VAM
fungi for inoculation of large tracts of land is not
presently available. However, it is important to
stress the potential significance of establishing VAM
islands as a source of VAM inoculum that is vital to
plant succession on mine spoil. The development of
practical and cost effective inoculation procedures
will advance the present revegetation process and
should ensure success of revegetated ecosystems in the
future.

LITERATURE CITED

1. Aldon, E. F. 1975. Endomycorrhizae enhance
    survival and growth of fourwing saltbush on coal
    mine spoils. USDA Forest Service Research Note
    RM-294:1-2.
2. Aldon, E. F. 1978. Endomycorrhizae enhance shrub
    growth and survival on mine spoils. Pages 174-179
    in: The Reclamation of Disturbed Arid Lands, R. A.
    Wright, ed., Univ. of New Mexico Press,
    Albuquerque
3. Allen, E. B. 1984, VA mycorrhizae and colonizing
    annuals: implications for growth, competition, and
    succession. Pages 41-51 in: VA Mycorrhizae and
    Reclamation of Arid and Semiarid Lands, S. E.
    Williams and M. F. Allen, eds., University of
    Wyoming Agriculture Experiment Station, SA 1261,
    Laramie.
4. Allen, E. B., and Allen, M. F. 1980. Natural
    re-establishment of vesicular-arbuscular
    mycorrhizae following stripmine reclamation in
    Wyoming. J. Appl. Ecol. 17:139-147.
5. Allen, E. B., and Allen, M. F. 1986. Water
    relations of xeric grasses in the field:
    interactions of mycorrhizas and competition. New

Phytol. 104:559-571.
6. Allen, E. B., and Allen, M. F. 1988. Facilitation of succession by the nonmycotrophic colonizer *Salsola kali* (Chenopodiaceae) on a harsh site: effects of mycorrhizal fungi. Amer. J. Bot. 75 (2):257-266.
7. Allen, M. F. 1984. Physiology of mycorrhizae: a key to understanding successful plant establishment. Pages 69-80 in: VA Mycorrhizae and Reclamation of Arid and Semiarid Lands, S. E. Williams and M. F. Allen, eds., University of Wyoming Agricultural Experiment Station, SA 1261, Laramie.
8. Allen, M. F. 1988. Belowground structure: a key to reconstructing a productive arid ecosystem. Pages 113-135 in: Reconstruction of Disturbed Arid Ecosystems, E. B. Allen, ed., Westview Press, Boulder, Colorado.
9. Allen, M. F. 1988. Belowground spatial patterning: Influence of root architecture, microorganisms and nutrients on plant survival in arid lands. Pages 113-135 in: The Reconstruction of Disturbed Arid Lands, E. B. Allen, ed., Westview Press, Boulder, Colorado.
10. Allen, M. F., Allen, E. B., and Friese, C. F. 1989. Responses of the non-mycotrophic plant *Salsola kali* to invasion by vesicular-arbuscular mycorrhizal fungi. New Phytol. 111:45-49.
11. Allen, M. F., and MacMahon, J. A. 1985. Impact of disturbance of cold desert fungi: Comparative microscale dispersion patterns. Pedobiologia 28:215-224.
12. Antonovics, J., Bradshaw, A. D., and Turner, R. G. 1971. Heavy metal tolerance in plants. Pages 1-85 in: Advances in Ecological Research, J.B. Cragg, ed., Academic Press, New York.
13. Barnhill, M. A. 1981. Endomycorrhizae in some nursery-produced trees and shrubs on a surface-mined area. Tree Planter's Notes 20-22.
14. Best, C. 1985. Vesicular-Arbuscular mycorrhizal associations in the revegetation of acid strip mine spoil. Pages 53-68 in: Fifth Better Reclamation with Trees Conference. C. A. Kolar,

ed., Carbondale, Illinois.
15. Bethlenfalvey, G. J., Franson, R. L., Brown, M. S., and Mihara, K. L. 1989. The *Glycine-Glomus-Bradyrhizobium* symbiosis. IX. Nutritional, morphological and physiological responses of nodulated soybean to geographic isolates of the mycorrhizal fungus *Glomus mosseae*. Physiol. Plant. 76:226-232.
16. Biondini, M. E., Bonham, C. D., and Redente, E. F. 1984. Relationships between induced successional patterns and soil biological activity of reclaimed areas. Reclamation and Revegetation Research 85:323-342.
17. Bosland, P. W., and Williams, P. W. 1987. An evaluation of *Fusarium oxysporum* from crucifers based on pathogenicity, isozyme polymorphism, vegetative compatibility, and geographic origin. Can. J. Bot. 65:2067-2073.
18. Brundett, M. 1991. Mycorrhizas in natural ecosystems. Advances Ecological Research 21:171-313.
19. Call, C. A., and McKell, C. M. 1982. Vesicular-arbuscular mycorrhizae - a natural revegetation strategy for disposed processed oil shale. Reclamation and Revegetation Research 1:337-347.
20. Call, C. A., and McKell, C. M. 1984. Field establishment of fourwing saltbush in processed oil shale and disturbed native soil as influenced by vesicular-arbuscular mycorrhizae. Great Basin Naturalist 44:363-371.
21. Chen, Y., and Aviad, T. 1990. The effects of humic substances on plant growth. Pages 161-185 in: Humic Substances in Soil and Crop Sciences: Selected Readings, P. MacCarthy, C. E. Clapp, R. L. Malcolm, and P. R. Bloom, eds., American Society of Agronomy - Soil Science Society of America.
22. Daft, M. J., and Nicolson, T. H. 1974. Arbuscular mycorrhizas in plants colonizing coal wastes in Scotland. New Phytol. 73:1129-1138.
23. Daniels, B. A., and Duff, D. M. 1978. Variation in germination and spore morphology among four

isolates of *Glomus mosseae*. Mycologia
70:1261-1267.

24. Danielson, R. M., Visser, S., and Parkinson, D.
1983. Microbial activity and mycorrhizal potential
of four overburden types used in the reclamation
of extracted oil sands. Can. J. Soil Sci.
63:363-375.

25. Danielson, R. M., Visser, S., and Parkinson, D.
1983. Plant growth in four overburden types used
in the reclamation of extracted oil sands. Can. J.
Soil Sci. 63:353-361.

26. Davies, F. T., and Call, C. A. 1987. Survival and
growth of mycorrhizal woody revegetation species
in Texas lignite overburden. Page 148 in: Proc.
7th North American Conf. on Mycorrhizae, D.
Sylvia, L. Hung and J. Graham, eds., Gainesville,
FL.

27. Elkins, N. Z., Parker, L. W., Aldon, E., and
Whitford, W. G. 1984. Responses of soil biota to
organic amendments in stripmine spoils in
northwestern New Mexico. J. Environ. Quality 13:
215-219.

28. Fresquez, P. R., Aldon, E. F., and Lindemann, W.
C. 1986. Microbial re-establishment and the
diversity of fungal genera in reclaimed coal mine
spoils and soils. Reclamation and Revegetation
Research 4:245-258.

29. Fresquez, P. R., and Lindemann, W. C. 1982. Soil
and rhizosphere microorganisms in amended coal
mine spoils. Soil Sci. Soc. Amer. J. 46:751-755.

30. Gildon, A., and Tinker, P. B. 1981. Interactions
of vesicular-arbuscular mycorrhizal infections and
heavy metals in plants. II. Effects of infections
on copper uptake. New Phytol. 95:263-268.

31. Gould, A. B., and Liberta, A. E. 1981. Effects of
topsoil storage during surface mining on the
viability of vesicular-arbuscular mycorrhiza.
Mycologia 73:914-922.

32. Janos, D. P. 1980. Mycorrhizae influence tropical
succession. Biotropica 12:56-64.

33. Johnson, N. C., and McGraw, A.-C. 1988.
Vesicular-arbuscular mycorrhizae in taconite
tailings. I. Incidence and spread of endogonaceous

fungi following reclamation. Agric., Ecosystems Environ. 21:135-142.

34. Johnson, N. C., and McGraw, A.-C. 1988. Vesicular-arbuscular mycorrhizae in taconite tailings. II. Effects of reclamation practices. Agric., Ecosystems Environ. 21:143-152.

35. Johnson, N. C., Zak, D. R., Tilman, D., and Pfleger, F.L. 1991. Dynamics of vesicular-arbuscular mycorrhizae during old field succession. Oecologia 86:349-358.

36. Khan, A. G. 1981. Growth responses of endomycorrhizal onions in unsterilized coal waste. New Phytol. 87:363-370.

37. Kiernan, J. M., Hendrix, J. W., and Maronek, D. M. 1983. Endomycorrhizal fungi occurring on orphan strip mines in Kentucky. Can. J. Bot. 61:1798-1803.

38. Killham, K., and Firestone, M. K. 1983. Vesicular arbuscular mycorrhizal mediation of grass response to acidic and heavy metal depositions. Plant Soil 72:39-48.

39. Koske, R. E. 1975. *Endogone* spores in Australian sand dunes. Can. J. Bot. 53:668-672.

40. Lambert, D. H., and Cole, H.,Jr. 1980. Effects of mycorrhizae on establishment and performance of forage species in mine spoil. Agron. J. 72:257-260.

41. Leisman, G. A. 1957. A vegetation and soil chronosequence on the Mesabi Iron Range spoil banks. Minnesota Ecological Monograph 27:221-245.

42. Leung, H., and Williams, P. H. 1986. Enzyme polymorphism and genetic differentiation among geographic isolates of the rice blast fungus. Phytopathology 76:778-783.

43. Lindemann, W. C., Lindsey, D. L., and Fresquez, P. R. 1984. Amendment of mine spoil to increase the number and activity of microorganisms. Soil Sci. Soc. Am. J. 48 (3):574-578.

44. Lindsey, D. L., Cress, W. A., and Aldon, E. F. 1977. The effects of endomycorrhizae on growth of rabbitbrush, fourwing saltbush, and corn in coal mine spoil material. USDA Forest Service Research Note RM-3431-6.

45. Loree, M. A. J., and Williams, S. E. 1984.
    Vesicular-arbuscular mycorrhizae and land
    disturbance. Pages 1-14 in: VA Mycorrhizae and
    Reclamation of Arid and Semiarid Lands, S. E.
    Williams and M. F. Allen, eds., University of
    Wyoming Agriculture Experiment Station, SA 1261,
    Laramie.
46. Loree, M. A. J., and Williams, S. E. 1987.
    Colonization of western wheat grass (*Agrpyron
    smithii* Rydb.) by vesicular-arbuscular mycorrhizal
    fungi during the revegetation of a surface mine.
    New Phytol. 106:735-744.
47. McGraw, A. C., and Schenck, N. C. 1980. Growth
    stimulation of citrus, ornamental and vegetable
    crops by select mycorrhizal fungi. Proc. Florida
    State Hort. Soc. 93:201-205.
48. McIlven, W. D., and Cole, H. 1976. Spore dispersal
    of Endogonaceae by worms, ants, and birds. Can. J.
    Bot. 54:1486-1489.
49. Miller, R. M. 1979. Some occurrences of
    vesicular-arbuscular mycorrhizae in natural and
    disturbed ecosystems of the Red Desert. Can. J.
    Bot. 57:619-623.
50. Miller, R. M, and Jastrow, J. D. 1992. The
    application of VA mycorrhizae to ecosystem
    restoration and reclamation. Pages 438-467 in:
    Mycorrhizal Functioning and Integrative Plant
    Fungal Process. M. F. Allen, ed., Chapman and
    Hall, New york. 534 pp.
51. Miller, R. M., and Jastrow, J. D. 1992. The role
    of mycorrhizal fungi in soil conservation. Pages
    29-44 in: Mycorrhizae in Sustainable Agriculture.
    G. J. Bethlenfalvay and R. G. Linderman, eds., ASA
    special publication 54. Madison, WI. 124 pp.
52. Morton, J. B., and Walker, C. 1984. *Glomus
    diaphanum*: A new species in the endogonaceae
    common in West Virginia. Mycotaxon 21:431-440.
53. Mosse, B. 1972. The influence of soil type and
    Endogone strain on the growth of mycorrhizal
    plants in phosphate deficient soil. Rev. Ecol.
    Biol. 9:529-537.
54. Mott, J. B., and Zuberer, D. A. 1987. Occurrence
    of vesicular-arbuscular mycorrhizae in mixed

overburden mine spoils of Texas. Reclamation and Revegetation Research 6:445-456.

55. Norland, M. R., Veith, D. L., and Dewar, S. W. 1991. The effects of compost age on the initial vegetation cover on coarse taconite tailing on Minnesota's Mesabi Iron Range. Pages 331-341 in: Proc. 8th Meet. Amer. Soc. Surf. Min. Reclam., Durango, CO.

56. Norland, M.R., Veith, D.L., and Dewar, S.W. 1991. Initial vegetative cover on coarse taconite tailing using organic amendments on Minnesota's Iron Range. Pages 263-277 in: Proc. 8th. Meet. Amer. Soc. Surf. Min. Reclam., Durango, CO.

57. Norland, M.R., Veith, D.L., and Dewar, S.W. 1992. Vegetation response to organic amendments on coarse taconite tailing. Pages 341-360 in: Proc. 9th. Meet. Amer. Soc. Surf. Min. Reclam., Duluth, MN.

58. Noyd, R.K., Pfleger, F.L., and Stewart, E.L. 1992. Native plants and vesicular-arbuscular mycorrhizal fungi in reclamation of coarse iron mine tailings in Minnesota. Pages 361-369 in: Proc.9th Meet. Amer. Soc. Surf. Min. Reclam., Duluth, MN.

59. Perry, D. A., Molina, R., and Amaranthus, M. P. 1987. Mycorrhizae, mycorrhizosphere, and reforestation: Current knowledge and research needs. Can. J. For. Res. 17:929-940.

60. Porter, W. M., Robson, A. D., and Abbott, L. K. 1987. Factors controlling the distribution of vesicular-arbuscular mycorrhizal fungi in relation to soil pH. J. Appl. Ecol. 24:663-672.

61. Qureshi, J. A., Thurman, D. A., Hardwick, K., and Collin, H.A. 1985. Uptake and accumulation of zinc, lead and copper in zinc and lead tolerant *Anthoxanthum odoratum L..* New Phytol. 100: 429-434.

62. Reeves, F. B., Wagner, D., Moorman, T., and Kiel, J. 1979. Role of endomycorrhizae in revegetation practices in the semi-arid West. I. A comparison of incidence of mycorrhizae in severely disturbed vs. natural environments. Amer. J. Bot. 66:1-13.

63. Rothwell, F. M. 1984. Aggregation of surface mine soil by interaction between VAM fungi and lignin

degradation products of lespedeza. Plant Soil
80:99-104.
64. Rothwell, F. M., and Eagleston, D. 1985. Microbial
relationships in surface-mine revegetation.
Environ. Geochem. Health 7:28-35.
65. Santos, P. F., and Whitford, W. G. 1981. The
effects of microarthropods on litter decomposition
in a Chihuahuan desert ecosystem. Ecology
62:654-663.
66. Schwab, S., and Reeves, F. B. 1981. The role of
endomycorrhizae in revegetation practices in the
semi-arid west. III. Vertical distribution of
vesicular-arbuscular (VA) mycorrhiza inoculum
potential. Amer. J. Bot. 68:1293-1297.
67. Shetron, S. G., and Duffek, R. 1970. Establishing
vegetation on iron mine tailings. J. Soil Water
Conserva. 25:227-230.
68. Stahl, P. D., and Smith, W. K. 1984. Effects of
different geographic isolates of *Glomus* on the
water relations of *Agropyron smithii*. Mycologia
76:261-267.
69. Stahl, P. D., Williams, S. E., and Christensen, M.
1988. Efficacy of native vesicular-arbuscular
mycorrhizal fungi after severe soil disturbance.
New Phytol. 110:347-354.
70. Stewart, E. L., and Pfleger, F. L. 1985. Selection
and utilization of mycorrhizal fungi in
revegetation of iron mining wastes. U.S.
Department of the Interior Final Project Report
1-64.
71. Stroo, H. F., and Jencks, E. M. 1982. Enzyme
activity and respiration in mine soils. Soil Sci.
Soc. Am. J. 46:548-553.
72. Sylvia, D. M. 1988. Growth response of native
woody plants to inoculation with *Glomus* species in
phosphate mine soil. Proc. Soil Crop Sci. Soc. Fl.
47:60-63.
73. Sylvia, D. M. 1990. Inoculation of native woody
plants with vesicular-arbuscular mycorrhizal fungi
for phosphate mine land reclamation. Agric.,
Ecosystems Environ. 31:252-261.
74. Sylvia, D. M., and Williams, S. E. 1992.
Vesicular-arbuscular mycorrhizae and environmental

stress. Pages 101-124 in: Mycorrhizae and Sustainable Agriculture. G. J. Bethlenfalvay and R. G. Linderman, eds., ASA special publication 54. Madison, WI. 124 pp.

75. Taber, R. A. 1982. Occurrence of *Glomus* spores in weed seeds in soil. Mycologia 74:515-520.

76. Tate, R. L., III, and Klein, D.A. 1985. Soil Reclamation Processes, Pages 1-349 in: Microbiological Analyses and Applications, R.L. Tate and D.A. Klein, eds., Marcel Dekkar, Inc., New York and Basel.

77. Visser, S. 1985. Management of microbial processes in surface mined land reclamation in western Canada. Pages 203-241 in: Soil Reclamation Processes: Microbiological Analyses and Applications, R. L. Tate III and D. A. Klein, eds., Marcel Dekkar,Inc., New York and Basel.

78. Vogel, W. G. 1982. A Guide for Revegetating Coal Minesoils in the Eastern United States. Pages 1-190 in: USDA For. Ser. Tech. Rep. NE-68, Northeast For. Exp. Stn., Broomhall, PA.

79. Vogel, W. G., and Rothwell, F. M. 1988. Mushroom compost and papermill sludge influence development of vegetation and endomycorrhizae on acid coal-mine spoils. Pages 206-213 in: Mine Drainage and Surface Mine Reclamation. Vol.II: Mine Reclamation, Abandoned Mine Lands and Policy Issues, U. S. Depart. Interior.

80. Waaland, M. E., and Allen, E. B. 1987. Relationships between VA mycorrhizal fungi and plant cover following surface mining in Wyoming. J. Range Management 40:271-276.

81. Wallace, P. M., and Best, G. R. 1983. Enhancing ecological succession: 6. Succession of vegetation, soils and mycorrhizal fungi following strip mining for phosphate. Pages 385-394 in: Surface Mining Hydrology, Sedimentology, and Reclamation, University of Kentucky, Lexington, KY.

82. Wallace, P. M., Best, G. R., and Feiertag, J. A. 1985. Mycorrhizae enhance growth of sweetgum in phosphate mined overburden soils. Pages 41-52 in: Better Reclamation with Trees Conference,

University of Southern Illinois, Carbondale, Il.
83. Warner, N. J., Allen, M. F., and MacMahon, J. A. 1987. Dispersal agents of vesicular-arbuscular mycorrhizal fungi in a disturbed arid ecosystem. Mycologia 79:721-730.
84. Zak, J. C., and Parkinson, D. 1982. Initial vesicular-arbuscular mycorrhizal development of slender wheatgrass on two amended mine spoils. Can. J. Bot. 60:2241-2248.
85. Zak, J. C., and Parkinson, D. 1983. Effects of surface emendation of two mine spoils in Alberta, Canada, on vesicular-arbuscular mycorrhizal development of slender wheatgrass: a 4-year study. Can. J. Bot. 61:798-803.

ROLE OF ECTOMYCORRHIZAL FUNGI IN MINESITE RECLAMATION

N. Malajczuk[1], P. Reddell[2] and M. Brundrett[3]

[1]CSIRO, Division of Forestry, Western Australia.
[2]CSIRO, Minesite Rehabilitation Research Program,
Queensland.
[3]Soil Science and Plant Nutrition Group, University of
Western Australia.

Mining of the landscape represents the extreme of soil and plant disturbance. A myriad of tight or tenuous links established between plants and soil microorganisms, which have evolved *in situ* over long periods of time, are broken. Restoration of these disturbed sites to functioning ecosystems requires an understanding of the many soil processes important in facilitating uptake, storage and cycling of nutrient and water, by the reintroduced plant species. Unfortunately, much of the functional diversity of soil microorganisms involved in those process is poorly understood.

Roughly 90 percent of plant species belong to families that form mycorrhizas; mutualistic associations between soil fungi and root tissue (4,21). Mycorrhizae improve seedling growth and survival by enhancing uptake of nutrients, increasing root longevity, and protecting against pathogens and soil stresses (7). There are basically two broad groups: those forming ectomycorrhizae (ECM), so termed because of the external modification to the root morphology. These are characterised by (a) the development of an external fungal sheath around the fine root tissue, and (b) the penetration of the fungus between root epidermis and cortex cells; and those termed vesicular-arbuscular mycorrhizae (VAM), the name coming from structures formed within the root cortical cells. Unlike the ECM, no external modification of the root accompanies VAM.

Trees forming ECM include most of the commercially important grown species in the temperate and boreal forests and 70 percent of the species planted in the tropics. Trees such as pines and oaks grown on sites that have not previously supported ECM die or do not grow well unless inoculated (14,15). For trees forming VAM, including the majority of tropical and many temperate deciduous families, the symbiosis is also important.

MYCORRHIZAL DIVERSITY AND SITE PRODUCTIVITY

There are thousands of fungal species that form ECM with plants. The type of ectomycorrhizae is attributed to different fungal symbionts and may vary with changes in the physical, chemical and biotic environment. Functional diversity of these fungi is potentially very high and it would appear that different fungi affect different hosts in different environments. Some mycorrhizal fungi may benefit the host plants in aiding nutrient uptake (3). Others benefit the hosts directly during periods of soil stresses at certain times during plant development, following disturbance, or indirectly by influencing important ecosystem properties such as soil structure. The diversity of mycorrhizal fungal associations is likely to contribute to the 'buffering capacity' of the forest ecosystem. Maintaining mycorrhizal fungal diversity helps minimise site degradation by increasing plant resilience to unpredictable or changing environments. Mycorrhizal fungal species are central to the concept of biodiversity in natural ecosystems and for successful tree species establishment. Because mycorrhizal fungi appear to have disproportionate influence on the survival and fitness of plants in new and restored habitats, we believe they are the cornerstone to the re-establishment of functioning ecosystems.

Incomplete knowledge of the abundance, diversity and distribution of soil organisms is particularly striking for some groups such as mycorrhizal fungi

associated with eucalypts. For example, until quite
recently, no large scale, systematic exploration of
mycorrhizal fungi had been described in Australia. In
recent expeditions to Australian tropical and
temperate rainforests in Queensland and Tasmania,
(Malajczuk, Trappe, Reddell, Bougher, Castellano,
Amaranthus and Young, unpublished) over 2000
mycorrhizal fungal specimens were collected
representing new families, genera and many new
species. Expeditions of this nature help to develop a
base for understanding the relationship of mycorrhizal
fungi and tree and forest function. However, there is
no simple relationship between biological diversity of
an ecosystem and its productivity. Nor is there a
simple relationship between loss of biological
diversity and productivity losses. These outcomes are
highly variable and likely dependent upon which
species and ecosystems are involved. As indicated
above, loss of mycorrhizal fungi from functioning
ecosystems would substantially decrease primary plant
productivity because they play a pivotal role in the
acquisition of nutrient, uptake of water and protect
roots against pathogens and soil stresses.

MYCORRHIZAL DYNAMICS

    In a forest environment, the diversity of
mycorrhizal fungal species is constantly changing.
Assessment of trees for their association with the
fruiting bodies (mushrooms, toadstools or truffles) of
ECM fungi show marked spatial and temporal
distribution, and similar patterns of distribution
have been observed for many different forest species
in both the temperate and tropical environments
(9,12). Studies by Mason (12) and Gardner and
Malajczuk (6) of birch and eucalypt stands,
respectively, indicated successional patterns occurred
in the distribution of sporocarps of specific fungal
genera. Further, not only do the dominant species of
mycorrhizal fungi change with stand age but the
species diversity increases with age of the forest
(9).
    Invariably these successional changes in the

fungal species is dynamically linked to the aboveground biota. Mycorrhizal fungi are often entirely dependent upon utilizing carbon directly from the roots as they do not decompose organic matter for their energy source. While some mycorrhizal fungi are not selective towards a particular plant species, many others form specific associations with particular plant species. For example, eucalypts form mycorrhizal associations with hundreds of fungi that are incompatible with other tree genera (16). Thus successional dynamics in forest communities can greatly influence below-ground biodiversity of mycorrhizal and vice versa. Mycorrhizal fungal diversity may also be closely tied to structural or habitat diversity in forest communities within the forest ecosystem. Ectomycorrhizal fungi predominate in the immediate vicinity of fine roots and the organic layers of soil (8,22). Trees or shrubs surviving on managed sites may contribute substantially to the maintenance of mycorrhizal fungal diversity. Similarly, many mycorrhizal fungi may require specific habitats such as litter on the forest floor. Habitat diversity may promote mycorrhizal diversity as conditions change within a growing season. In periods of adequate soil moisture, organic matter supports the highest level of ectomycorrhizal activity but during periods of drought or following removal of litter or top soil, deeper soil profiles harbour the greater proportion of surviving fungi (20).

In the past few decades, the extent of temperate and tropical forest has been reduced dramatically due to ever-increasing demand for wood fibre and requirements for agricultural land. Because above and below-ground biota are tightly linked, such changes are likely to result in dramatic losses in genetic diversity. Hopes for possible restoration of degraded sites are dependent on detailed knowledge of the site, as well as access to gene pools from which to select desired species (both tree and mycorrhizal fungus) for replanting.

THE IMPACT OF DISTURBANCE ON ECTOMYCORRHIZAL FUNGI

Soil disturbance can have a severe impact on the propagules that are responsible for the survival and spread of ECM fungi in soil (Figure 1). In undisturbed natural communities, a network of fine hyphae and mycelial strands or rhizomorphs attached to existing mycorrhizae is believed to be primarily responsible for the spread of fungi to new roots (4,17). However, we would expect severe soil disturbance to eliminate most of these hyphae and the efficacy of any surviving portions of the hyphal network to be further curtailed by the absence of host plants to provide energy. Propagules that, at least initially, can maintain their inoculum potential in disturbed soils are thought to include spores, mycorrhizal fragments, surviving segments of rhizomorphs and sclerotia (2,4). We would expect fungi which do not produce resistant propagules (sclerotia, etc.) to be lost immediately following disturbance, while others would decline due to attrition of their propagules during subsequent soil storage (Figure 1). In natural undisturbed communities, ectomycorrhizal fungi are dispersed by airborne spores that are seasonally produced in vast numbers, or by the activity of animals which feed on subterranean basidiocarps and ascocarps that contain spores (11,13). In a disturbed habitat, the effectiveness of these natural vectors will be dependent on the quality and proximity of undisturbed habitats containing suitable fungi (and their associated animal vectors) as well as the seasonality of the fruiting of these fungi (19).

In undisturbed habitats, a wide diversity of fungi can form ECM associations with one plant species, but the diversity of these fungi in disturbed habitats is typically much lower and there are some species which are characteristic of disturbed sites (5,6,12). These reductions in fungal diversity could result because disturbance events themselves eliminate many intolerant fungi, while others with more resistant propagules could still be ineffective because they had a limited capacity to adapt to the

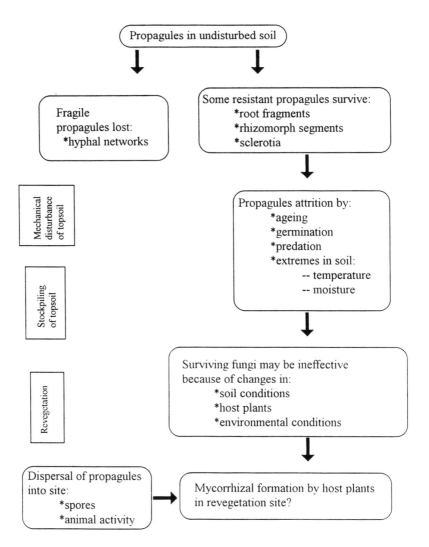

Figure 1. Probable impacts of disturbance on ectomycorrhizal fungus propagules.

major changes to the environment (4). Typical changes that occur when soil is disturbed and stockpiled include the loss of organic matter, nutrients, soil aggregate structure and biological diversity (1,5). There are also many cases where the substrate to be revegetated (such as waste rock dump material) would initially be devoid of soil organisms, as is

considered in the case study presented later in this
chapter.
    The probability of a host plant becoming
mycorrhizal in a disturbed habitat depends on the
survival or dispersal of propagules of fungi which can
adapt to prevailing site conditions (Figure 1).
Figure 2 outlines the steps that should be taken to
determine if the introduction of additional
mycorrhizal fungus inoculum is appropriate and the
methodology required. The questions (Figure 2) can be

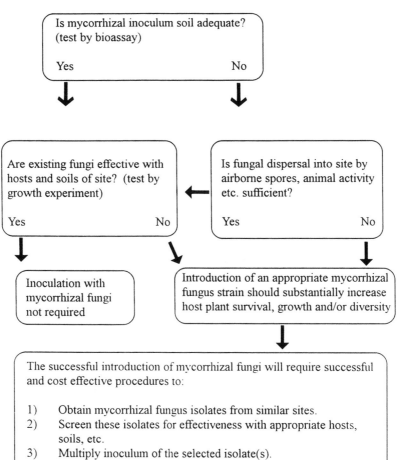

Figure 2. Evaluating the potential benefit of mycorrhizal introduction.

answered by experiments such as bioassays, where host plants are grown in soil from particular sites and inoculation trials are conducted to measure mycorrhizal benefits in these soils. For example, bioassay results from a study in the Alligator Rivers region where uranium is mined indicated that ECM fungus inoculum was generally absent from minesite areas with sparse vegetation, but were present in most soil samples from woodland areas or disturbed sites where host trees or shrubs had become established (Figure 3). Methodology for selecting isolates of ECM fungi and producing inoculum is outlined (Figure 2) and is described in greater detail in Chapter 11.

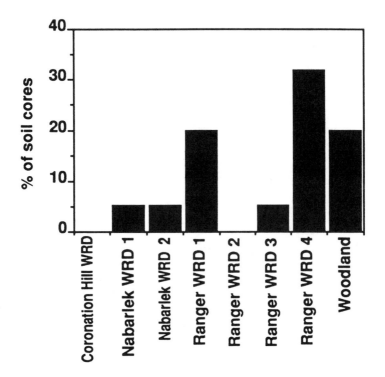

Figure 3. Percent of soil cores with ectomycorrhizal fungal inoculum collected from Alligator Rivers Region sites

CASE STUDY: THE ROLE OF ECTOMYCORRHIZAL IN REVEGETATION OF WASTE ROCK DUMPS AT A URANIUM MINE IN TROPICAL NORTHERN AUSTRALIA

Rehabilitation without Topsoil

Revegetation strategies at many minesites involve the use of topsoils to cover mine floors, waste rock dumps, and clay and rock layers that cap tailings dams and other waste materials. Topsoils are used because often (a) their physical and chemical characteristics make them suitable media for plant establishment and growth, and (b) they provide a reservoir of seed and soil biota, including mycorrhizal fungi, that aid in reclamation of the site. However, in some situations this 'traditional' rehabilitation approach is economically and/or environmentally impractical due to the location and nature of the mining operations and the limited quantity of topsoil available. In addition, prolonged storage of topsoil prior to use can substantially reduce its value as an 'inoculum' of soil biota and as a source of seed and other regenerative plant propagules (Figure 1). An alternative approach to rehabilitation is presented that eliminates the need to store and spread topsoils, but requires the development of techniques to re-introduce soil biota, including ECM fungi, to ensure the success of the revegetation process.

This research was conducted at the Ranger Uranium Mine in the monsoonal tropics of the Northern Territory of Australia. The aim of rehabilitation at this site was to produce a self-sustaining community of local woodland species on an engineered land-surface formed from waste rock. A range of rudimentary soils had been observed to form *in situ* from weathering of waste-rock on the dumps at Ranger Mine within a time-span of about 5 years in the humid tropical environment of the region (18). Examination of the morphological, chemical, physical, and mineralogical properties of these soils demonstrated that they had many features in common with natural undisturbed soils in the surrounding landscape, but were distinguishable in terms of their higher pH, much lower organic carbon content, and higher

concentrations of some plant nutrients. These
properties suggest that the 'minesites' could support
plant growth without the addition of topsoils in the
rehabilitation process. However, a potentially
critical limitation to vegetation establishment on
these rudimentary soils was the absence of soil fauna
and microflora, including mycorrhizal fungi (3,10).
Strategies to introduce soil biota, and especially
mycorrhizal fungi, into revegetation sites on these
'mine soils' was identified as the focus of a research
program that commenced in 1988.

Ectomycorrhizal Fungi in Revegetation without Topsoil

As a starting point to determine the importance
of mycorrhizae in revegetation at Ranger Mine, a
survey was conducted to establish the mycorrhizal
status of plants in natural woodland communities of
the area. The majority of plant species in the forest
and woodland communities surrounding the mine were
found to be mycorrhizal (Figure 4a). Ectomycorrhizae
occurred in 32% of species from 5 families (Myrtaceae,
Caesalpiniaceae, Fabaceae, and Minosaceae). These ECM
fungal species include most of the dominant trees of
the woodland communities and they comprise most of the
woody biomass (Figure 4b). Consequently, ECM fungi
are an integral component of the native woodlands and
it is likely that the tree flora of the area are
highly dependent on their ECM fungal associations for
both establishment and efficient nutrient acquisition.

Following on from this baseline survey, studies
were conducted on soils from these waste rock dumps,
from the surrounding native forest, and from
stockpiled natural topsoils. These studies assessed:
1. comparative diversity of sporocarps of presumed
    ECM fungi in the native woodlands and on the
    oldest mine areas that had been rehabilitated
    using topsoil,
2. ECM fungal infectivity in soils from (a) the
    native woodlands, (b) stockpiled topsoils of
    varying ages, (c) areas on the waste rock dumps
    revegetated 5 to 7 years previously using
    topsoil, and (d) bare soils forming *in situ* on
    the dumps from the weathering of waste rock, and

3. the role of inoculation with ECM fungi in overcoming nutritional limitations to tree growth in a young 'mine soil' forming *in situ* from weathering waste rock.

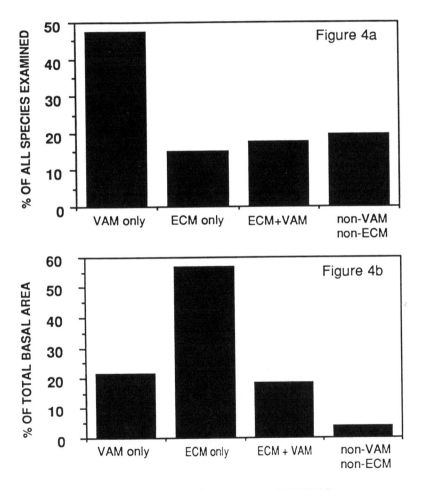

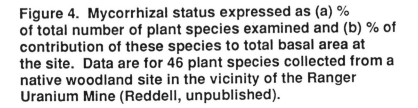

Figure 4. Mycorrhizal status expressed as (a) % of total number of plant species examined and (b) % of contribution of these species to total basal area at the site. Data are for 46 plant species collected from a native woodland site in the vicinity of the Ranger Uranium Mine (Reddell, unpublished).

Mine areas revegetated 5 to 7 years previously using topsoil had a depauperate flora of ECM fungi incomparison to the native woodlands as has been found in other studies (6). Sporocarps of only 2 genera of ECM fungi were found on the rehabilitated areas compared to 23 in the woodland areas (Table 1). Ectomycorrhizal colonization of *Eucalyptus porrecta* Blake in soils varied substantially (Table 2). Ectomycorrhizas were present in all three woodland soils, but were absent or less infective in stockpiles of natural soils (irrespective of the age of the stockpile) and in soils from areas on the waste rock dumps revegetated 5 to 7 years previously using topsoil and tubestock. The limited diversity of ECM fungi on both the vegetated and bare mine areas probably reflects both (a) the successional sequence of fungi that are adapted to the new sites and (b) the poor effectiveness of transport of propagules of many ECM fungi to these sites in applied topsoil and by natural agencies.

A glasshouse experiment identified that a typical, young soil developing from weathering waste rock was deficient in P and Zn for optimum growth of *E. pellita* Blake. Inoculation with ECM fungi partially alleviated these deficiencies and resulted in significant increases in plant growth compared to the uninoculated treatment (Table 3). However, although inoculation with ECM fungi partially overcame the particular nutrient limitations, judicious use of some fertilizer in plant establishment on these 'mine soils' may be necessary to optimise the functioning of the mycorrhizal symbioses.

Implications for ecosystem development and practical applications

The research presented in this case study provides an example of the approach and type of information that are required to (i) assess the importance of ectomycorrhizas in restoration of mined or other degraded land and (ii) evaluate the potential benefits of mycorrhizal fungal introduction to accelerating the rehabilitation process as outlined in

TABLE 1.   Systematic arrangement and fruiting characteristics of presumed ectomycorrhizal fungi collected in woodlands of the Kakadu area and on waste rock dumps at Ranger Uranium Mine (Reddell, unpublished).

Order and Family	Genus	Fruiting habit[1]	Collected from: native woodlands	areas revegetated with topsoil
BASIDIOMYCETES				
Hymenomycetes				
AGARICALES				
Amanitaceae	*Amanita*	e	+	-
Boletaceae	*Boletellus*	e	+	-
	Boletus	e	+	-
	Tylopilus	e	+	-
	"*Rhizopogon*-like"	h	+	-
Cortinariaceae	*Cortinarius*	e	+	-
	Inocybe	e	+	-
Russulaceae	*Lactarius*	e	+	-
	Russula	e	+	-
	Zelleromyces	h	+	-
Tricholomataceae	*Laccaria*	e	+	-
	Tricholoma	e	+	-
APHYLLOPHORALES				
Cantharellaceae	*Cantharellus*	e	+	-
Clavariaceae	*Ramaria*	e	+	-
CHONDROGASTRALES				
Chondrogastraceae	*Chondrogaster*	h	+	-
Gasteromycetes				
PHALLALES				
Hysterangiaceae	*Hysterangium*	h	+	-
PODAXALES				
Cribbeaceae	*Cribbea*	h	+	-
SCLERODERMATALES				
Sclerodermataceae	*Horakiella*	h	+	-
	Pisolithus	e	+	+
	Scleroderma	e	+	+
Taxa of uncertain affinities				
undescribed genus 1		h	+	-
undescribed genus 2		h	+	-
ZYGOMYCETES				
ENDOGONALES				
Endogonaceae	*Endogone*	h	+	-

[1]sporocarps are: e = epigeous (fruit aboveground)
              h = hypogeous (fruit belowground)

TABLE 2.   Ectomycorrhizal colonization of *Eucalyptus porrecta* seedlings grown for 6 and 12 weeks under glasshouse conditions, in unamended soils collected from the Ranger mine area (adapted from Reddell and Milnes, 1992).

SITE	Ectomycorrhizal colonization (% root length colonized) at:	
	6 weeks	12 weeks
Rudimentary soils forming *in situ* from waste rock		
no vegetation	0	0
no vegetation	0	0
no vegetation	0	0
Soil from waste rock dump areas revegetated using topsoils		
5 year old revegetation, unburnt	0	0
5 year old revegetation, burnt	6	23
7 year old revegetation	0	0
Stockpiled topsoils from the native woodland		
1 year stockpiled	0	10
4 year stockpiled	0	0
8 year stockpiled	10	28
Undisturbed, native woodland soils		
low open woodland	25	44
woodland	12	36
tall open forest	23	53

TABLE 3.   Role of ectomycorrhizal fungi in alleviating P and Zn deficiency in *Eucalyptus pellita* seedlings growing on young soils forming *in situ* on waste rock dumps at Ranger (20 weeks after planting). Values are for means ± standard deviation (Adapted from Reddell and Milnes, 1992).

TREATMENT	Shoot dry weight (g/plant)	Ectomycorrhizal colonization (%)
(a) Phosphorus		
Complete nutrients minus P, no inoculation	0.01 ± 0.01	0
Complete nutrients minus P, inoculated *Scleroderma* spores	0.25 ± 0.07	8 ± 8
Complete nutrients minus P, inoculated *Pisolithus* spores	0.59 ± 0.12	33 ± 19
(b) Zinc		
Complete nutrients minus Zn, no inoculation	0.13 ± 0.04	0
Complete nutrients minus Zn, inoculated *Scleroderma* spores	2.46 ± 1.11	43 ± 18
(c) Controls		
No nutrients added, no inoculation	0.01 ± 0.01	0
Complete nutrients, no inoculation	4.94 ± 1.07	0
No nutrients added, no inoculation	0.01 ± 0.01	0
Complete nutrients, no inoculation	4.94 ± 1.07	0

Figure 2. The results show that the diversity and infectivity of ECM fungi on areas rehabilitated up to 8 years previously with topsoil is low. This indicates that natural dispersal and re-establishment of ECM fungi on the waste rock dumps occurs at a very slow rate, even when topsoil has been used. This may impact significantly on the rate of development and resilience to environmental stresses of the developing plant community. Their low numbers and diversity may also influence the sustainability of any response to applied nutrients. As a consequence, the potential benefits of inoculation with ECM fungi to the general 'health' of the restored plant community are high.

In practical terms, rehabilitation strategies implemented at this minesite using soils forming *in situ* on the waste rock dumps will be needed to ensure deficiencies of nutrients are alleviated and that viable populations of ECM fungi are introduced and maintained during early phases of vegetation establishment. Research currently in progress is investigating methods for introducing ECM fungi into these soils and determining their effects on the plant growth and subsequent community development.

CONCLUSION

The restoration of degraded sites with tree species is seen as a challenge to identify and re-introduce populations of soil micro-organisms which are essential in the functioning of ecosystems. Ectomycorrhizas, which aid the plant in nutrient and water uptake and assist in protecting roots from soil stresses, are seen as a key component of this restoration process. However, our limited knowledge of the ecology, diversity and dynamics of ECM fungi in many ecosystems is a major impediment to developing a more predictive understanding of requirements for, and responses to, ECM inoculation of plant communities being re-established on mined lands.

LITERATURE CITED

1. Abdul-Kareem, A. W., and McRae, S. G. 1984. The

effects of topsoil of long-term storage in
stockpiles. Plant Soil 76:357-363.

2. Ba, M. A., Garbaye, J., and Dexheimer, J. 1991.
Influence of fungal propagules during the early
stage of the time sequence of ectomycorrhizal
colonization of *Afzellia africana* seedlings.
Can. J. Bot. 69:2442-2447.

3. Bougher, N. L., Grove, T. S., and Malajczuk, N.
1990. Growth and phosphorus acquisition of karri
(*Eucalyptus diversicolor* F. Muell) seedlings
inoculated with ectomycorrhizal fungi in relation
to phosphorus supply. New Phytol. 114:77-85.

4. Brundrett, M. C. 1991. Mycorrhizas in natural
ecosystems. Pages 171-313 in: Advances in
Ecological Research. Vol. 21. A. Macfayden, M.
Begon, and A. H. Fitter, eds. Academic Press,
London.

5. Danielson, R. M. 1985. Mycorrhizae and
reclamation of stressed terrestrial environments.
Pages 173-201 in: Soil Reclamation Processes -
microorganisms, analyses and applications'. R.
L. Tate and D. A. Klein, eds. Marcel Dekker, New
York.

6. Gardner, J. H., and Malajczuk, N. 1988.
Recolonization of rehabilitated bauxite mine
sites in Western Australia by mycorrhizal fungi.
For. Ecol. Management 24:27-42.

7. Harley, J. L., and Smith, S. E. 1983.
Mycorrhizal Symbiosis. Academic Press, New York.

8. Harvey, A. E., Jurgensen, M.F., and Larsen, M.J.
1979. Clearcut harvesting and ectomycorrhizae:
survival of activity on residual roots and
influence on a bordering forest stand in western
Montana. Can. J. For. Res. 10:300-303.

9. Hilton, R. N., Malajczuk, N., and Pearce, M. H.
1988. Large fungi of the jarrah forest: an
ecological and taxonomic survey. Pages 89-110 in:
The Jarrah Forest. A complex Mediterranean
ecosystem. B. Dell, J. Havel and N. Malajczuk,
eds. Kluwer Press, Amsterdam.

10. Janos, D. P. 1980. Mycorrhizae influence
tropical succession. Biotropica 12(2, Suppl):
56-64

11. Malajczuk, N., Trappe, J. M., and Molina, R. 1987. Interrelationships amongst some ectomycorrhizal trees, hypogeous fungi and small mammals: Western Australian and northwestern American parallels. Austr. J. Ecol. 12:53-55.

12. Mason, P. A., Wilson, J., Last, F. T., and Walker, C. 1983. The concept of succession in relation to the spread of sheathing mycorrhizal fungi on inoculated tree seedlings growing in unsterile soil. Plant Soil 71:247-256.

13. Maser, C., and Maser, Z. 1988. Interactions amongst squirrels, mycorrhizal fungi and coniferous forests in Oregon. Great Basin Naturalist 48:358-369.

14. Mikola, P. 1970. Mycorrhizal inoculation in afforestation. International Review of Forestry Research 3:123-196.

15. Mikola, P. 1973. Application of mycorrhizal symbiosis in forestry practice. Pages 383-406 in: Ectomycorrhizae-their ecology and physiology, G. C. Marks and T. T. Kozlowski, eds. Academic Press, New York.

16. Molina, R., Massicote, H., and Trappe, J. M. 1992. Specificity phenomena in mycorrhizal symbioses: community-ecological consequences and practical implication. Pages 357-423 in: Mycorrhizal Functioning; an integrative plant-fungus process, M. F. Allen, ed., Chapman and Hall, New York.

17. Ogawa, M. 1985. Ecological characters of ectomycorrhizal fungi and their mycorrhizae-an introduction to the ecology of higher fungi. JARC 18:305-314.

18. Reddell, P., and Milnes, A. R. 1992. Mycorrhizas and other specialized nutrient-acquisition strategies: their occurrence in woodland plants from Kakadu and their role in rehabilitation of waste rock dumps at a local uranium mine. Austr. J. Bot. 40:223-242.

19. Reddell, P., and Spain, A. V. 1991. Earthworms as vectors of viable propagules of mycorrhizal fungi. Soil Biol. Biochem. 23:767-774.

20. Reddell, P., and Malajczuk, N. 1984. Formation

of mycorrhizae by jarrah (*Eucalyptus marginata* Donn ex Smith) in litter and soil. Austr. J. Bot. 32:511-520.

21. Trappe, J. M. 1987. Phylogenetic and ecological aspects of mycotrophy in the Angiosperms from an evolutionary standpoint. Pages 5-26. in: Ecophysiology of VA mycorrhizal plants, G. R. Safir, ed., CRC Press, Boca Raton, FL.

22. Trappe, J. M., and Fogel, R. D. 1977. Ecosystematic functions of mycorrhizae. Pages 205-214 in: The belowground ecosystem: a synthesis of plant associated processes. Rangelands Science Department Science Series 26.

THE EFFECTS OF CULTURAL PRACTICES AND PESTICIDES ON VAM FUNGI

James E. Kurle and F. L. Pfleger
University of Minnesota
St. Paul, Minnesota

Modern agriculture faces a dilemma. High input crop production practices yield an abundance of food, oil, and fiber products. The same practices result in a troubled farm economy and serious environmental problems, including soil erosion and water pollution with fertilizers, pesticides, and animal wastes. Public opinion and economics compel farmers and researchers to consider alternative agricultural practices "which reduce costs and protect environmental quality by enhancing beneficial biological interactions and natural processes" (53). Among these biological interactions, the symbiosis formed by specific fungi and plant roots called vesicular arbuscular mycorrhizae (VAM) has the potential for maintaining plant vigor while reducing the need for chemical and fertilizer inputs. Management of the VAM symbiosis for maximum crop plant benefit provides both an opportunity and a challenge for agricultural scientists.

The potential importance of VAM to plant health has been discussed in reviews by Harley and Smith (26), Powell and Bagyaraj (65), Safir (67), and Sieverding (74). VAM fungal species and even strains within a species can vary greatly in their associations with plants ranging from mutualistic to parasitic. Their beneficial role in phosphorus, nitrogen, and micronutrient uptake has been demonstrated for many crops. VAM fungus colonization also has been associated with increased tolerance to water stress and decreased susceptibility to a variety of plant diseases. A less frequently recognized role of VAM is the hyphal contribution to soil aggregation and soil structure (84). The challenge is to incorporate VAM into agricultural management practices with predictable, reliable and beneficial effects on plant health.

Numerous factors hinder the application of VAM in agricultural production. Approximately 150 species of VAM fungi are recognized on the basis of spore morphology. Within each species are isolates with unique growth requirements depending on the soils, climate, and host where the isolate was obtained. The unique behavior of each species and isolate is complicated by interactions with host plants, soil type, soil microflora, climate, and agricultural practices. The complexity of the interactions may appear intimidating, however, a body of applied studies has been completed which allow generalizations about the effects of agricultural practices on VAM in field situations.

Agricultural practices may alter VAM fungal populations, species composition, and colonization. The relationship of these factors to plant productivity is not always apparent. The occurrence of VAM may be expressed in terms of spore density, root colonization percentage, or in terms of infectivity using measures such as mycorrhizal inoculum potential (MIP) or most probable number (MPM) (64,50). Spore quantity alone is an inadequate measure of the potential of a soil to establish mutualism between fungus and host plant since it measures the sporulating capacity of the VAM fungus and says little about the viability of the spores or the importance of other forms of inoculum. The relationship of VAM fungus spore number to host plant colonization is also unclear, especially in mixed populations of VAM fungi. Furthermore, the relationship of colonization to crop yields is ambiguous (47). Colonization percentage and infectivity assays offer no information about the species composition of the VAM fungal community which can only be inferred from the presence and identification of spores. Since intraradical and extraradical hyphae cannot be easily identified as to species, the fungal species competitiveness and effectiveness is poorly understood in field situations. The use of monoclonal antibodies (87) offers the possibility of *in situ* species identification, however, antibodies exist for only a limited number of species. What is needed is the ability to relate species, root colonization percentage, and hyphal development to measures of biological productivity and nutrient use efficiency such as yield, biomass production, and nutrient uptake. With the limitations of these measures in mind, we can discuss successful examples of VAM management in agricultural production. These examples suggest that VAM are already being managed, sometimes inadvertently, to benefit plant health.

In this chapter we examine the effects of cultural practices on VAM, emphasizing information gathered from field studies. We will also consider VAM responses to tillage, crop sequence, plant breeding, and fertilizer and pesticide application. The effects of flooding, burning, and compaction will be discussed briefly.

TILLAGE

In field studies, the intensity of tillage may affect VAM fungal colonization and the efficiency of nutrient uptake in early plant development. Vivekanaden and Fixen (88) found higher VAM colonization and large differences in early growth responses to P when ridge-tilled corn (*Zea mays* L.) or soybeans (*Glycine max* (L.) Merrill) was compared to corn or barley (*Hordeum vulgare* L.) which had been moldboard plowed (Figure 1). A similar reduction in colonization of winter wheat occurred when mechanically tilled fallow was compared to a no-tillchemical fallow system (90). VAM formation and plant dry weight of young soybeans decreased as the amount of secondary tillage and field traffic was increased (52). Corn also showed higher VAM formation, higher P uptake, and higher final yield under no-till than under conventional tillage (1).

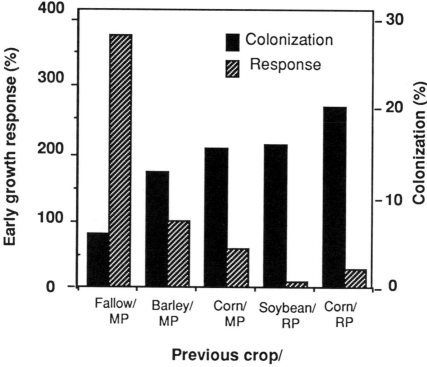

Figure 1. The relationship of previous crop, tillage, VAM colonization, and early growth response (V6 vegetative stage) to added P. Colonization is reduced after fallow periods and moldboard plowing (MP). When a reduced tillage system, ridge planting (RP), is used, colonization is increased. Higher colonization percentages are associated with improved P nutrition and a reduced requirement for added P. Early growth response is calculated as: (Dry weight of unfertilized plant - dry weight of fertilized plant)/(Dry weight of unfertilized plant) x 100. (Modified from Vivekananden and Fixen, 1991).

The capacity for P uptake by VAM is related to the extent of extraradical hypha formation (35). The most important effect of tillage on VAM function is

disruption of the extraradical hyphae.   In greenhouse
studies which helped to explain the effect of tillage,
O'Halloran et al. (60) found that corn grown under no-
till management had higher P uptake than corn grown
under conventional tillage.   Follow-up studies
(17,18,19,20,48) demonstrated that reduced P uptake
was a consequence of disruption of the VAM fungal
hyphal network.   Although VAM fungal inoculum consists
of spores, hyphae and infected roots, the intact
hyphal network appears to be most important in
mycorrhizal fungus colonization of roots.
Colonization early in plant growth was accelerated by
the presence of preexisting hyphae, so tillage-caused
disturbance reduced colonization.   The preexisting
hyphal network was also effective in P transport as
soon as colonization occurred and reduced the carbon
expenditure required to reestablish a hyphal network.
As a result, P uptake and dry matter production was
enhanced by VAM fungal colonization early in plant
development even at relatively high levels of soil P
application.(18,38,39).

Compaction related to tillage and other field
activities may also interfere with VAM fungal
colonization.   Increased bulk density resulting from
field traffic significantly reduced VAM formation but
increased spore number (52,86).   Moisture stress and
anaerobic soil conditions resulting from reduced soil
porosity may also have reduced mycorrhizal activity by
limiting photosynthate available to the fungus from
the plant.

Although we are not aware of any published
research relating tillage with shifts in VAM fungal
species composition, hypothetically such relationships
exist.   Where more intensive tillage systems are used,
resulting in greater soil disturbance, it is likely
that VAM fungal species which sporulate more heavily
would be favored.   Species which are more dependent on
the mycelial network and on hyphal remnants in root
fragments for carry-over would be disadvantaged
because of disruption of the network and accelerated
decomposition of roots.   Seasonal pattern of VAM
fungal growth and sporulation would be important in
determining the multiplication of a particular fungal

species. Those species which sporulate most intensely early in the growing season would be favored in many crops, since post-harvest tillage would interfere with the sporulation of species which colonize and multiply later in the growing season.

FLOODING AND BURNING

Flooding may reduce VAM fungal inoculum potential. In studies of various rice (*Oryza sativa* L.), corn, legume crop sequences, rice-rice sequences reduced VAM fungal propagule numbers, possibly as a result of low soil oxygen levels (33,68). Burning, which is practiced to eliminate crop residues, control diseases, or to clear land, may also reduce VAM fungal colonization. Burning can affect VAM fungi directly through heating of the soil surface, or indirectly through germination inhibitors produced in soils by heating (89).

FERTILIZERS

The effects of fertilizer or lime addition on VAM are well documented in greenhouse studies (67). These studies have shown that the fertilizer effects on colonization and spore production are influenced by initial soil fertility, soil type, organic matter status, and host plant and endophyte species. Studies carried out in the field have demonstrated the complexity of interactions among these factors as well as environmental and seasonal influences.

The factor which most strongly influences VAM fungal colonization and sporulation is the P status of the plant. This in turn, determines cell membrane permeability and root exudation of carbohydrates and amino acids available as metabolites to the fungus (22). In the field, the addition of P fertilizer to soils reduces either VAM fungal colonization or sporulation in a variety of crops including corn, soybeans, clover (*Trifolium* spp.) and small grains (1,25,30,69). Seemingly contradictory studies have also shown increased colonization with the addition of P (5,23). The response of VAM to P addition appears

to be an interaction of native soil fertility, P
addition, and plant health. VAM formation at very low
soil P levels is limited by impaired plant
productivity. As soil P is increased by
fertilization, plant P status improves and VAM
formation increases (5).

Soil N and K interact with P to influence
colonization by VAM fungi. Nitrogen deficiency in
onion (*Allium cepa* L.) resulted in reduced
colonization (81). The application of high rates of N
(200 kg/ha) to rotations of wheat (*Triticum aestivum*
L.), barley and green manure crops reduced VAM fungus
inoculum potential (3,4). Strzemska (79) reported a
similar reduction in VAM formation at high N
application rates on corn. The application of
fertilizer containing P and K and either no N or
excessive N also resulted in decreased colonization
when compared to more equally proportioned
applications of the three minerals, while a low rate
of K addition was associated with the highest VAM
colonization (23). This is similar to the observation
of Plenchette and Corpron (63) that low rates of K
were associated with high levels of colonization by
VAM fungi. In contrast, the addition of complete
fertilizers did not affect VAM formation (69).

When chemical fertilization was combined with the
addition of manures, the results were more complex.
Manure alone increased spore numbers and root
colonization, while increasing rates of a fertilizer
containing proportional rates of N, P, and K did not
change colonization by VAM fungi. However, the
combination of manure and increasing rates of N, P, or
K alone resulted in reduced VAM fungal inoculum
potential (25). The varied results of combined
fertilizer and manure application may reflect
differences in composition and quality of manures,
since the effect of manure alone on VAM formation has
been shown to depend on the manure source, addition
rate, nutrient content, and possibly its state of
decomposition (8,74).

VAM fungal species differ in their response to
chemical fertilizers. *Sclerocystis* spp. disappeared
from tropical soils when they were fertilized and

cultivated (74). In greenhouse studies, *Glomus intraradices* (Schenck & Smith) was unaffected by P fertilization, while sporulation and colonization by *Acaulospora longula* (Spain & Schenck) and *Gigaspora margarita* (Becker & Hall) was suppressed (15,83). Nitrogen and P interacted to influence root colonization by *G. etunicatum* (Becker & Gerdemann) and *G. margarita* (81). The differential response of VAM fungal species to P fertilization may result in selection of species less sensitive to fertilization and less effective as mutualists (40). Isolates of a single species can also exhibit a differential response to P fertilization. For example, tropical isolates of *G. clarum* (Nicholson & Schenck) from low fertility soils were more effective in enhancing plant growth in low P soils than isolates taken from high fertility soils (45). It appears that intensive agricultural practices involving high fertilizer application rates may reduce the efficiency of P uptake by limiting VAM fungal colonization of crop plants. In some cases, fertilization may select VAM fungal species or isolates that are tolerant of high fertilizer rates but contribute little to nutrient uptake.

CROP SEQUENCE

VAM fungi are obligate symbionts and are profoundly affected by changes in host plant species and populations. In agricultural systems, these changes usually occur annually and result in the complete elimination of one plant species and its replacement by another. In some cases, crop sequences, especially those which include fallow periods may have marked negative as well as positive effects on VAM fungal populations.

Fallow periods are often included in crop sequences to control weeds and/or pathogens, or conserve moisture. Tillage for weed control during this period may range from multiple cultivations to no-till chemical fallows. In the absence of a host crop, VAM fungal inoculum potential decreases as hyphae and spores are parasitized and spores continue to germinate. Tillage causes a further decrease in inoculum potential (Figure 1). In a wheat-fallow rotation, inoculum potential decreased 80% in plowed fallow areas, but only 25% in areas where weeds were controlled with herbicides and tillage was omitted. In low fertility soils in Australia, extended periods of fallowing resulted in reduced crop yield (Figure 2). Inoculation of such fallow soils with soil which had recently supported a crop eliminated the effects of the long fallow period. The extended fallow period was accompanied by a decline in VAM fungal inoculum potential (82).

Although species of most plant families are colonized by mycorrhizal fungi, members of the Brassicaceae and Chenopodiaceae are considered to be nonhosts. Prior cropping with nonhosts has been shown to reduce VAM fungal inoculum potential. Spore numbers in soil following a crop of mustard (*Brassica juncea* (L.) Czern & Coss) were similar to those found in fallowed areas and were 16% lower than those found in areas where cowpeas (*Vigna unguiculata* (L.) Walp) were grown. Root colonization of cowpea planted after mustard was 14% lower than root colonization of cowpea planted after cowpea (25). Black and Tinker (7) reported a 45% reduction in colonization of barley (*Hordeum distichon* L.) planted following kale (*Brassica oleraceae* L.) compared to barley following barley. When a host and nonhost crop were grown together inoculum potential was actually increased because of more extensive hyphal development (59). In addition, Ocampo (58) observed in a greenhouse study that cabbage (*B. oleraceae* L.), a nonhost, did not cause reduced colonization when grown in rotation with a host crop. Because the planting of mustard as a cover or "smother" crop is being advocated as an alternative to herbicides for weed control, the effect of intercropping host and nonhost crops deserves further attention in field studies.

The use of host crops as cover crops to increase VAM fungal inoculum potential appears to be a practical alternative to the application of inoculum. Pre-cropping with cassava (*Manihot esculenta* Crantz), kudzu (*Pueraria phaseoloides* Bonth.) or sorghum

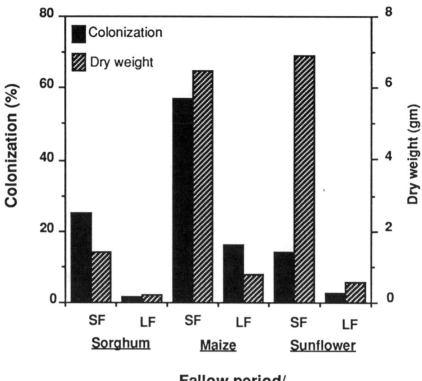

Figure 2. The relationship of the length of fallow period, VAM colonization, and plant dry weight in early plant development (three to six weeks). Long fallow (LF) periods (14 to 20 weeks) resulted in lower colonization percentages and were associated with reduced plant dry weight. Short fallow (SF) periods (four to eight weeks) resulted in higher colonization percentages and increased plant dry weight. (Modified from Thompson, 1987).

(*Sorghum* sp.) increased VAM fungal spore numbers, early season VAM formation and final yields of cowpea and *Stylosanthes capitata* Vog. on land recently converted to cultivation from native vegetation (13).

The dynamics of colonization and sporulation differ among fungal species and isolates of species. These characteristics are influenced by edaphic and environmental factors interacting with the fungus and the host plant. If their functioning were more fully understood, they could possibly be manipulated to favor more beneficial species or isolates as an alternative to inoculation.

In a greenhouse study, sporulation of *G. fasciculatum* (Thaxter sensu (ed) Gerd. & Trappe) was strongly influenced by host plant species, while *G. macrocarpum* (Tul. & Tul.) Gerd. & Trappe and *G. mosseae* (Nicholson & Gerd.) Gerd. & Trappe were not significantly affected. Sporulation of all three VAM fungal species was heaviest on sudangrass (*Sorghum vulgare*) var. sudanense (Piper) Hitch) when compared to asparagus (*Asparagus officinalis* L.), marigold (*Tagetes erecta* (L.), red clover (*Trifolium pratense* L.) or tomato (*Lycopersicon esculentum* Mill). Sporulation was poorly correlated with colonization percentage but strongly correlated with plant dry matter production. Strubbles and Skipper (78) observed that bahiagrass (*Paspalum notatum* Fliigge), corn and sudangrass were all superior to soybeans as host plants for spore production. They suggested that the rapidly growing root system of the grasses provided for more rapid colonization and sporulation. Differences in colonizing ability existed among both mycorrhizal fungal species and isolates within species when sudangrass, little bluestem (*Schizachyrium scoparium* (Michx.) Nash), big bluestem (*Andropogon gerardii* Vitman), and prairie dropseed (*Sporobolus heterolepis* (Gray) Gray) were used as hosts (11). This preferential colonization might result in populations of VAM fungi selected by the presence of particular host plants (28).

Similar shifts in VAM fungal species composition and populations resulting from crop sequence have been observed in field studies. In a study examining the effect of crop sequence in a tropical multiple cropping system, higher numbers of spores and infective propagules were present after cropping with peanut (*Arachis hypogea* L.) than after finger millet (*Eleusine coracana* (L.) Gaerth) (25). The increased mycorrhizal activity appeared to improve growth of the subsequent crop of sunflowers (*Helianthus annuus* L.). Monocultures of six crops; corn, sorghum, bahia grass, soybean, cotton (*Gossypium hirsutum* L.), and peanut, which replaced native perennial vegetation in a temperate ecosystem, resulted in unique VAM fungal communities under each crop. Higher spore numbers occurred in association with soybeans than any of the other crops. In addition, spore number was highly correlated with colonization percentage which differed significantly among the six crops. There was also a change in species dominance (70). A similar shift in species dominance and diversity occurred when a tropical savannah was converted to agricultural production (71).

The shift in VAM fungal species composition which occurs in agroecosystems may result in selection of species or isolates which are more tolerant of agricultural management practices but less beneficial to crop plants. Johnson et al. (42) suggested that such a shift in species occurred under monocultures of corn or soybeans. Yield and tissue mineral composition of corn were negatively correlated with spore numbers of VAM fungal species which appeared to proliferate in corn (Figure 3) (43). A similar relationship existed with VAM fungal species proliferating in association with soybeans. It was hypothesized that continuous cropping selected VAM fungal species which grew and sporulated most rapidly. Those species were not beneficial, and as their numbers increased, crop vigor declined (43).

Weeds are an important component of any agroecosystem and may have a major influence on VAM fungal species and populations. Pellet and Sieverding (62) observed increased spore numbers of *Glomus* spp. when either *Euphorbia hirta* L. or *Digitaria sanguinalis* (L.) Scop. were the dominant weeds in plantings of beans (*Phaseolus vulgaris* L.) or casava. They suggested that weeds might be tolerated as alternate hosts for mycorrhizal fungi if the benefit to subsequent crops outweighed yield losses resulting

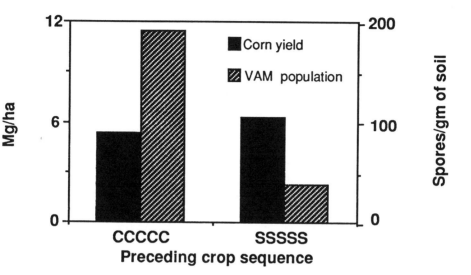

Figure 3. Relationship of crop sequence, crop yield, and VAM populations. The population of *Glomus occultum* appears to increase when corn is grown continuously and is negatively correlated with corn yield. (Modified from Johnson et al., 1992).

from weed competition. Increases in spore numbers were also seen in both conventional and low input management systems with substantial annual grass weed populations (42,44) or with combinations of weeds and cover crops (14). Crops grown in sustainable cropping systems might even benefit from the presence of weeds as hosts for VAM fungal species (21). However, if ineffective mycorrhizal fungus species are favored by the weed species present in a particular crop or area, the presence of weeds would interfere with the maintenance of populations of effective indigenous or introduced VAM fungal species.

BREEDING

It is possible that the crop itself could be made especially responsive to VAM. Cultivars of soybean (27), corn (85), and alfalfa (*Medicago sativa* L.) (57) vary in levels of dependency on VAM. Wheat cultivars show differing degrees of VAM colonization and dependency (2) and differ in their responsiveness to VAM (46) and different VAM fungal isolates. When modern high-yielding varieties are compared with landraces and primitive wheat ancestors, the modern varieties may show little growth response or even negative growth response to VAM formation (Figure 4). The studies suggest the possibility of increasing the P use-efficiency of crop plants by selection for VAM responsiveness, since the differences among cultivars appear to have a genetic basis which can be manipulated using traditional plant breeding methods (46). The results suggest also that crop breeding programs may be inadvertently selecting for genotypes which are unresponsive to VAM, since colonization percentage was similar among landraces and high yielding varieties.

PESTICIDES

The use of chemical control agents is a fundamental component of modern high-input, large scale agricultural production. While integrated pest management systems will reduce the use of some chemicals, the use of others may be increased by practices such as minimum tillage systems or herbicide resistant cultivars. Information presently available is largely empirical and has little foundation in VAM physiology or ecology. Thus the effect of agricultural chemicals on VAM is poorly understood, and pesticide application may have inadvertent or unrecognized effects. In some instances, pesticides used with the intention of promoting plant health may impair or eliminate VAM activity, and thus damage plant health. Unfortunately observations of pesticide effects on VAM fungi are still often limited to measures of sporulation or root colonization. Much of

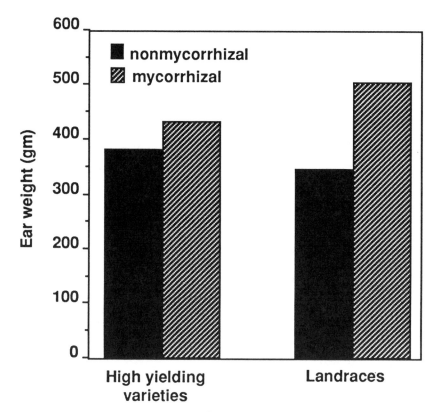

Figure 4.   Comparison of the effect of colonization by the VAM species *Glomus manihotis* on ear dry weight of 22 high yielding varieties and 22 landraces when grown in low P fertility conditions. Yields of landraces increased 46% compared to a 12% yield increase by the high yielding varieties. (Modified from Manske, 1990).

the current research has been conducted in greenhouse or growth chambers, in sterile media, or in media whose biological, physical or chemical characteristics have little similarity to field conditions. Here we review the response of VAM to some of the more commonly used fumigants, fungicides, herbicides, insecticides, and nematicides. We attempt to emphasize field studies and use greenhouse or growth chamber studies to explain results found in field studies.

Soil Fumigants

Methyl bromide-chloropicrin and vapam, two fumigants available for general use, are normally applied to orchard, nursery, and other horticultural crops. Conventional practices are based on the assumption that elimination of pathogens would result in stimulation of plant growth. However, fumigation has caused stunting in a range of crops including celery, onion and pepper (24), cotton (32), soybean (66), and corn (41). This reduction in plant productivity is often caused by decreased VAM formation which results in poor nutrient uptake. However, field studies suggest that the elimination of other soil microorganisms may also be involved in the effect (31).

Fungicides

Fungicides include an enormous variety of compounds which differ in their effect on host physiology, mode of action, method of application and formulation. As a result, it is difficult to generalize about a class of compounds or even a single compound. Since many of the fungicides reviewed in this section are applied as foliar sprays and are not systemic, it is questionable whether the amount of chemical reaching the soil would have a detrimental effect on mycorrhizal fungi.

The dicarboximide, captan, is typically applied as a foliar spray, seed treatment, or soil drench. It has produced a range of effects on VAM formation depending on VAM fungal species and host plant. It stimulated colonization of beans by *Glomus* spp. (10), had no effect on undetermined species infecting onion (16), and reduced *G. mosseae* colonization of corn (80). When applied at a high rate to sour orange, captan reduced colonization but not sporulation by *G. etunicatum* Becker & Gerd. (54). Application at half the recommended rate to bent grass (*Cloris gayana* Kunth) resulted in greater root colonization and spore numbers then the untreated control (77).

The dithiocarbamate fungicides mancozeb, maneb, thiram, and zineb are used to control many different plant pathogens. However, little information exists about their effect on specific VAM fungus species (37,77,80). In general, their effect on VAM fungi consisted of reduction in colonization and spore number. Similarly, the aromatic hydrocarbons, botran, chloroneb, chlorthalonil, lanstan and quintozene usually reduced VAM (16,49,56,80). Quintozene severely reduced VAM and P uptake (36). The effect of chloroneb depended on VAM fungal species involved (76).

Systemic fungicides are of particular interest because of their persistence within the plant, activity on organisms competing with VAM fungi for infection sites, or possible direct effects on VAM function. Application of the carboxin fungicide, vitavax, to sorghum seed significantly reduced root colonization by *G. deserticola* Trappe, Bloss. & Menge (51) early in plant development. In many plants, P needs are high during this period and inhibition of root colonization by VAM fungi may contribute to a P deficiency. In contrast, the systemic fungicides fosetyl-Al and metalaxyl enhanced VAM formation and plant growth. These two fungicides are specific in their control of oomycetous plant pathogens with no effect on other fungal groups including the Zygomycetes (73). It has been suggested that reductions in populations of organisms antagonistic to VAM fungi (29) and an increase in the availability of soluble sugars in the host provides an environment which is more favorable for root colonization and spread of VAM fungi (34,72).

Sterol-inhibiting fungicides are a group of chemicals which are becoming increasingly important because of their effectiveness against leaf pathogens. Although they can be applied in much lower quantities than fungicides applied as protectants, they are systemic and persistent within the plant. They have been shown to affect VAM fungal colonization and sporulation to differing degrees. Tridemorph was shown to increase (12) and decrease (37) root colonization by VAM fungi depending on plant species and time of application. Propiconazole (Tilt) applied to big bluestem inoculated with *G. etunicatum* did not appear to affect VAM formation. However, P uptake was reduced 20 fold compared to uptake by the inoculated but untreated control (Figure 5) (31). The results demonstrate the potential effects of this class of fungicide on VAM effectiveness. They also demonstrate the importance of examining VAM fungal characteristics other then colonization and sporulation when determining the effect of a pesticide.

Herbicides

Herbicides are designed to impede the growth of "weedy" higher plants. However, it is possible that they may directly affect VAM density and effectiveness by interference with mycorrhizal physiological processes or indirectly by changes in host plant physiology or populations. Alachlor applied at normal rates to soybeans did not appear to have a direct effect on root growth and VAM fungal colonization of soybeans. When applied at high rates, however, both soybean root growth and VAM fungus hyphal elongation were reduced. The effect persisted after the roots had recovered from inhibition, indicating a direct effect on the fungus (9). Difenzoquat inhibited spore germination of *G. etunicatum*, *G. mosseae*, and *G. monosporum* Gerdemann & Trappe (55). Simazine stimulated VAM fungal colonization of a nonhost plant quinoa, *Chenopodium quinona* Willd., by increasing root exudates (72). Sieverding and Liehner (75) observed both direct and indirect effects of herbicides on VAM fungal species and population colonizing cassava. In both greenhouse and field studies, VAM fungal spore numbers and level of root colonization were reduced by the use of diuron plus alachlor, oxadiazon, and oxifluorfen. Oxadiazon selectively reduced spore numbers of *Glomus* species in a greenhouse study and in the field. The results were not uniform, however, and

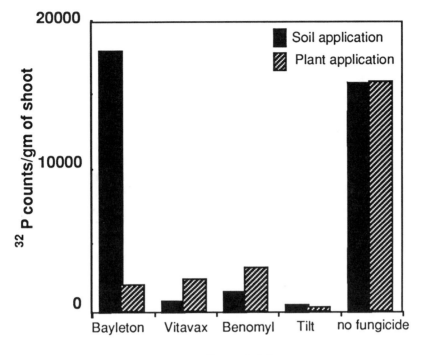

Figure 5. The effect of the application of five fungicides on P uptake by big bluestem (*Andropogon gerardii*) inoculated with *Glomus etunicatum*. The effect of the fungicide was strongly influenced by the site of application. There was also little relationship between colonization percentage and P uptake since the application of Bayleton to the soil and Tilt to either the plant or the soil resulted in the same colonizaton as no fungicide application. The application of Bayleton also resulted in greater P uptake than when no fungicide application was made. (Modified from Hetrick et al., 1988).

depended on soil type. Part of the reduction in spore numbers was attributed to a reduction in host weed populations caused by herbicide use. Similarly, in a Minnesota study (44), spore numbers were lower in areas of a corn-soybean rotation where management current or past practices reduced weed populations (44).

Nematicides and Insecticides

Organophosphate insecticides and nematicides such as carbaryl and diazinon had no adverse effect on root colonization by VAM fungi (9, 61). The carbamate insecticides carbofuran and aldicarb increased VAM formation on a variety of crops, possibly by limiting populations of fungivorous nematodes or other soil arthropods which feed on VAM fungal hyphae (74,77). The nematicide DBCP (1,2-dibromo-3-chloropropane) increased colonization of cotton roots by an unidentified VAM fungal species (6) also by reducing populations of fungivorous arthropods and nematodes. As with other pesticides, too little information is available to make generalizations about the effects of insecticides or nematicides on VAM.

The variety of VAM responses to pesticides reflects the multiplicity of pesticides used in farming. The responses also are a product of the procedures used to test the compounds and may not indicate the consequences of their use in agriculture. In some cases, the results suggest that pesticides intended to improve plant health may actually diminish it because of their effects on VAM. However, until the development of a standardized set of testing protocols which reflect actual pesticide use in the field, effects of agricultural chemicals on VAM will remain unclear.

CONCLUSIONS

VAM play a significant and complex role in plant health. The symbiosis with crop plants may increase yields by limiting the effects of abiotic disease such as nutrient deficiency or moisture stress. Either directly or indirectly, VAM also reduce susceptibility to biotic disease.   In contrast, VAM may impair plant health if the symbiosis becomes parasitic.

Many cultural practices used in high input, mechanized farming affect VAM fungi.   Some practices, such as crop rotation, minimum tillage and the use of cover crops, improve crop vigor by positively affecting VAM formation.   Other cultural practices which could be developed or modified to exploit the VAM symbiosis include:

1) Breeding plants for either increased colonization or increased responsiveness to VAM to improve P use efficiency or reduce disease susceptibility.
2) Managing weed communities or planting cover crops to increase VAM fungal inoculum potential especially during fallow periods.   A corollary of this approach is an examination of the effect of nonhost smother crops on VAM effectiveness.
3) Modification of P recommendations to account for increased P use efficiency resulting from VAM.

Economic and social pressures dictate that maximum agricultural productivity be achieved at minimum economic and environmental cost.   Agricultural systems which enhance naturally occurring biological phenomena, such as VAM, are a means of achieving those multiple objectives.   Field and laboratory research have demonstrated that cultural practices have profound effects on VAM and could be effective tools for their management.   If these fungi can be reliably and predictably managed, the mycorrhizal symbiosis will be a major tool for improving plant health in sustainable agricultural systems.

LITERATURE CITED

1. Anderson, E. L., Millner, P. D., and Kunishi, H. M. 1987. Maize root length density and mycorrhizal infection as influenced by tillage and soil phosphorus. J. Plant Nutrition 10:1349-1356.
2. Azcon, R., and Ocampo, J. A. 1981. Factors affecting the vesicular-arbuscular infection and mycorrhizal dependency of thirteen wheat

cultivars. New Phytol. 87:677-685.

3. Baltruschat, H., and Dehne, H. W. 1988. The occurrence of vesicular-arbuscular mycorrhiza in agro-ecosystems. I. Influence of nitrogen fertilization and green manure in continuous monoculture and in crop rotation on the inoculum potential of winter wheat. Plant Soil 107:279-284.

4. Baltruschat, H., and Dehne, H. W. 1989. The occurrence of vesicular-arbuscular mycorrhiza in agro-ecosystems: II. Influence of nitrogen fertilization and green manure in continuous monoculture and in crop rotation on the inoculum potential of winter barley. Plant Soil 113:251-256.

5. Bethlenfalvay, G. J., Bayne, H. G., and Pacovsky, R. S. 1983. Parasitic and mutualistic associations between a mycorrhizal fungus and soybean: The effect of phosphorus on host plant-endophyte interactions. Physiol. Plant 57:543-548.

6. Bird, G. W., Rich, J. R., and Glover, S. U. 1974. Increased endomycorrhizae of cotton roots in soil treated with nematicides. Phytopathology 64:48-51.

7. Black, R., and Tinker, P. B. 1979. The development of endomycorrhizal root systems. II. Effect of agronomic factors and soil conditions on the development of vesicular-arbuscular mycorrhizal infection in barley and on the endophyte spore density. New Phytol. 83:401-413.

8. Brechelt, A., 1987. Wirkung verschiedener organischer Dungemittel auf die Effizienz der VA-Mykorrhiza. J. Agronomy Crop Sci. 158:287-288.

9. Burpee, L., and Cole, H. Jr. 1978. The influence of alachlor, trifluralin, and diazinon on the development of endogenous mycorrhizae in soybeans. Bull. Environ. Contam. Toxicol. 19:191-197

10. De Bertoldi, M., Giovannetti, M., Griselli, M., and Rambelli, A. 1977. Effects of soil application of Benomyl and Captan on the growth of onions and the occurrence of endophytic mycorrhizae and rhizosphere microbes. Annals Appl. Biology 86:111-115.

11. Dhillion, S. S. 1992. Evidence for host-mycorrhizal preference in native grassland

species. Mycol. Res. 96:359-362.

12. Dodd, J. C., and Jeffries, P. 1989. Effect of fungicides on three vesicular-arbuscular mycorrhizal fungi associated with winter wheat (*Triticum aestivum* L.). Biol. Fertil. Soils 7:120-128.

13. Dodd, J. C., Arias, I., Koomen, I., and Hayman, D. S. 1990. The management of populations of vesicular-arbuscular mycorrhizal fungi in acid-infertile soils of a savanna ecosystem: II. The effects of pre-crops on the spore populations of native and introduced VAM-fungi. Plant Soil 122:241-247.

14. Douds, D. D., Jr., Janke, R. R., and Peters, S. E. 1990. Effect of conventional and low-input sustainable agriculture upon VAM fungi. Pages 88-89 in: Eighth North American Conf. on Mycorrhizae. M. Allen and S. Williams, eds., Jackson, WY. 5-8 September 1990. Univ. of Wyoming Agric. Exp. Stn., Laramie, WY.

15. Douds, D. D. Jr., and Schenck, N. C. 1990. Relationship of colonization and sporulation by VA mycorrhizal fungi to plant nutrient and carbohydrate contents. New Phytol. 116:621-627.

16. El-Giahami, A. A., Nicolson, T. H. and Daft, M. J. 1976. Effects of fungal toxicants on mycorrhizal maize. Trans. Brit. Mycol. Soc. 67:172-173.

17. Evans, D. G., and Miller, M. H. 1988. Vesicular-arbuscular mycorrhizas and the soil-disturbance-induced reduction of nutrient absorption in maize. I. Causal relations. New Phytol. 110:67-75.

18. Evans, D. G., and Miller, M. H. 1990. The role of the external mycelial network in the effect of soil disturbance upon vesicular-arbuscular mycorrhizal colonization of maize. New Phytol. 114:65-71.

19. Fairchild, G. L., and Miller, M. H. 1988. Vesicular-arbuscular mycorrhizas and the soil-disturbance-induced reduction of nutrient absorption in maize. II. Development of the effect. New Phytol. 110:75-84.

20. Fairchild, G. L. , and Miller, M. H. 1990. Vesicular-arbuscular mycorrhizas and the soil-

disturbance-induced reduction of nutrient absorption in maize. III. Influence of P amendments to soil. New Phytol. 114:641-650.

21. Gonzales-Chavez, M. C., Ferrera-Cerrato, R., and Garcia, R. 1990. The taxonomy of VA mycorrhizae in a sustainable agro-ecosystem in the humid tropics of Mexico. Page 120 in: Eighth North American Conference on Mycorrhizae. M. Allen and S. Williams, eds., Jackson, WY 5-8 September 1990. Univ. of Wyoming Agric. Exp. Stn., Laramie, WY.

22. Graham, J. H., Leonard, R. T., and Menge, J. A. 1981. Membrane-mediated decrease in root exudation responsible for phosphorus inhibition of vesicular-arbuscular mycorrhiza formation. Plant Physiol. 68:548-552.

23. Gryndler, M., Lestina, J., Moravec, V., Prikryl, Z., and Lipavsky, J. 1989. Colonization of maize roots by VAM-fungi under conditions of long-term fertilization of varying intensity. Agric., Ecosystems Environ. 29:183-186.

24. Haas, J. H., Bar-yosef, B., Krikun, J., Barak, R., Markovitz, T., and Kramer, S. 1987. Vesicular-arbuscular mycorrhizal fungus infestation and phosphorus fertigation to overcome pepper stunting after methyl bromide fumigation. Agron. J. 79:905-910.

25. Harinikumar, K. M., and Bagyaraj, D. J. 1989. Effect of cropping sequence, fertilizers, and farmyard manure on vesicular-arbuscular mycorrhizal fungi in different crops over three consecutive seasons. Biol. Fertil. Soils 7:173-175.

26. Harley, J. L., and Smith, S. E. 1983. Mycorrhizal Symbiosis. Academic Press. New York. 483 pp.

27. Heckman, J. R., and Angle, J. S. 1987. Variation between soybean cultivars in vesicular-arbuscular mycorrhiza fungi colonization. Agron. J. 79:428-430.

28. Hetrick, B. A. D., and Bloom, J. 1986. The influence of host plant on production and colonization ability of vesicular-arbuscular mycorrhizal spores. Mycologia 78:32-36.

29. Hetrick, B. A. D., and Wilson, G. W. T. 1991.

Effects of mycorrhizal fungus species and metalaxyl application on microbial suppression of mycorrhizal symbiosis. Mycologia 83:97-102.

30. Hetrick, B. A. D., Hetrick, J. A., and Bloom, J. 1984. Interaction of mycorrhizal infection, phosphorus level, and moisture stress in growth of field corn. Can. J. Bot. 62:2267-2271.

31. Hetrick, B. A. D., Wilson, G. T., Kitt, D. G., and Schwab, A. P. 1988. Effects of soil microorganisms on mycorrhizal contribution to growth of big bluestem grass in non-fertile soil. Soil Biol. Biochem. 20:501-507.

32. Hurlimann, J. H. 1974. Response of cotton and corn growth in fumigated soil. Ph. D. Dissertation, Univ. Calif., Berkeley, 94 pp.

33. Ilag, L. L., Rosales, A. M., Elazegui, F. A. and Mew, T. W. 1987. Changes in the population of infective endomycorrhizal fungi in a rice-based cropping system. Plant Soil 103:67-73.

34. Jabaji-Hare, S. H., and Kendrick, W. B. 1985. Effects of fosetyl-Al on root exudation and on composition of extracts of mycorrhizal and nonmycorrhizal leek roots. Can. J. Plant Pathol. 7:118-126.

35. Jakobsen, I., Abbott, L. K., and Robson, A. D. 1992. External hyphae of vesicular-arbuscular mycorrhizal fungi associated with Trifolium subterraneum L. I. Spread of hyphae and phosphorus inflow into roots. New Phytol. 120:371-380.

36. Jalali, B. L. 1978. Response of soil fungitoxicants on the development of VA mycorrhiza and phosphate uptake in cereals. Abst.Page 182 in: 3rd International Congress of Plant Pathology, Munchen.

37. Jalali, B. L., and Domsch, K. H. 1975. Effect of systemic fungitoxicants on the development of endotrophic mycorrhizae. Pages 619-626 in: Endomycorrhizae. F. E. Sanders, B. Mosse, P. B. Tinker, eds. Academic Press. London.

38. Jasper, D. A., Abbott, L. K., and Robson, A. D. 1989. Soil disturbance reduces the infectivity of external hyphae of vesicular-arbuscular mycorrhizal fungi. New Phytol. 112:93-99.

39. Jasper, D. A., Abbott, L. K., and Robson, A. D. 1989. Hyphae of a vesicular-arbuscular mycorrhizal fungus maintain infectivity in dry soil, except when the soil is disturbed. New Phytol. 112:101-107.

40. Jasper, D. A., Robson, A. D., and Abbott, L. K. 1979. Phosphorus and the formation of vesicular-arbuscular mycorrhizas. Soil Biol. Biochem. 11:501-505.

41. Jawson, M. D., and Aiken, R. M. 1988. Corn response to soil fumigation within crop rotations. Page 218 in: Abst. American Society of Agronomy Annual Meetings. 27 November-2 December. Anaheim, CA.

42. Johnson, N. C., Pfleger, F. L., Crookston, R. K., Simmons, S. R. and Copeland, P. J. 1991. Vesicular-arbuscular mycorrhizas respond to corn and soybean cropping history. New Phytol. 117:657-663.

43. Johnson, N. C., Copeland, P. J., Crookston, R. K., and Pfleger, F. L. 1992. Mycorrhizae: Possible explanation for yield decline with continuous corn and soybean. Agron. J. 84:387-390.

44. Kurle, J. E., and Pfleger, F. L. 1993. The effect of management and edaphic factors on vesicular-arbuscular mycorrhizal populations in a corn-soybean rotation. Agron. J. In Press.

45. Louis, I., and Lim, G. 1988. Differential response in growth and mycorrhizal colonization of soybean to inoculation with two isolates of Glomus clarum in soils of different P availability. Plant Soil 112:37-43.

46. Manske, G. G. B. 1990. Genetical analysis of the efficiency of VA mycorrhiza with spring wheat. Agric., Ecosystems Environ. 29:273-280.

47. McGonigle, T. P. 1988. A numerical analysis of published field trials with vesicular-arbuscular mycorrhizal fungi. Functional Ecology 2:473-478.

48. McGonigle, T. P., Evans, D. G., and Miller, M. H. 1990. Effect of degree of soil disturbance on mycorrhizal colonization and phosphorus absorption by maize in growth chamber and field experiments. New Phytol. 116:629-636.

49. Menge, J. A., Johnson, E. L. V. and Minassian, V. 1979. Effect of heat treatment and three pesticides upon the growth and reproduction of the mycorrhizal fungus *Glomus fasciculatus*. New Phytol. 82:473-480.

50. Moorman, T., and Reeves, R. B. 1979. The role of endomycorrhizae in revegetation practices in the semi-arid west. II. A bioassay to determine the effect of land disturbance on mycorrhizal populations. Amer. J. Bot. 66:14-18.

51. Mott, J., and Odvody, G. 1987. Effects of seed treatment chemicals on vesicular-arbuscular mycorrhizae. Page 32 in: Mycorrhizae in the next decade:Practical applications and research priorities. Proc. Seventh North American Conf. on Mycorrhizae. D.M. Sylvia, L.L. Hung, and J.H. Graham eds., Gainesville, FL 3-8 May 1987. Univ. Florida, Gainesville.

52. Mulligan, M. F., Smucker, A. J. M., and Safir, G. F. 1985. Tillage modifications of dry edible bean root colonization by VAM fungi. Agron. J. 77:140-144.

53. National Research Council 1989. Alternative Agriculture. National Academy Press. Washington, D. C. 448 pp.

54. Nemec, S. 1980. Effects of 11 fungicides on endomycorrhizal development on sour orange. Can. J. Bot. 58:522-526.

55. Nemec, S., and Tucker, D. 1983. Effects of herbicides on endomycorrhizal fungi in Florida citrus (*Citrus* spp.) soils. Weed Sci. 31:417-431.

56. Nesheim, O. N., and Linn, M. B. 1969. Deleterious effect of certain fungitoxicants on the formation of mycorrhizae on corn by *Endogone fasciculata* and on corn root development. Phytopathology 59:297-300.

57. O'bannon, J. H., Evans, D. W., and Peaden, R. N. 1980. Alfalfa varietal response to seven isolates of vesicular-arbuscular mycorrhizal fungi. Can. J. Plant Sci. 60:859-863.

58. Ocampo, J. A., 1980. Effect of crop rotations involving host and non-host plants on vesicular-arbuscular mycorrhizal infection of host plants.

Plant Soil 56:283-291.
59. Ocampo, J. A., Martin, J., and Hayman, D. S. 1980. Influence of plant interactions of vesicular-arbuscular mycorrhizal infections. I. Host and non-host plants grown together. New Phytol. 84:27-35.
60. O'Halloran, I. P., Miller, M. H., and Arnold, G. 1986. Absorption of P by corn (*Zea mays* L.) as influenced by soil disturbance. Can. J. Soil Sci. 66:287-302.
61. Parvathi, K., Venkateswarlu, K. and Rao, A. S. 1985. Toxicity of soil-applied fungicides to the vesicular-arbuscular mycorrhizal fungus *Glomus mosseae* in groundnut. Can. J. Bot. 63:1673-1675.
62. Pellet, D., and Sieverding, E. 1986. Host preferential multiplication of fungal species of the endogonaceae in the field, demonstrated with weeds. Pages 555-557 in: Physiological and genetical aspects of mycorrhizae. Mycorrhizae: Physiology and Genetics. V. Gianinazzi-Pearson and S. Gianinazzi, eds., Dijon, 1985. INRA. Paris.
63. Plenchette, C., and Corpron, I. 1987. Influence of P and K fertilization on VA mycorrhizal fungi population. Page 35 in: Mycorrhizae in the next decade practical applications and research priorities. Proc. Seventh North American Conference on Mycorrhizae. D. M. Sylvia, L. L. Hung, and J. H. Graham, eds., Gainesville, FL 3-8 May, 1987. Univ. FL Gainesville.
64. Porter, W. M. 1979. The 'most probable number' method for enumerating infective propagules of vesicular arbuscular mycorrhizal fungi in soil. Aust. J. Soil. Res. 17:515-519.
65. Powell, C. Ll., and Bagyaraj, D. J. eds., 1984. VA Mycorrhiza. CRC Press. Boca Raton, FL. 234 pp.
66. Ross, J. P., and Harper, J. A. 1970. Effect of Endogone mycorrhiza on soybean yields. Phytopathology 60:1552-1556.
67. Safir, G. R. 1987. ed., Ecophysiology of VA mycorrhizal plants. CRC Press. Boca Raton, rL. 224 pp.
68. Saif, S. R. 1983. The influence of soil aeration on the efficiency of vesicular-arbuscular

mycorrhizas. II. Effect of soil oxygen on growth and mineral uptake in *Eupatorium odoratum* L., *Sorghum bicolor* (L.) Moench and *Guizotia abyssinica* (L.f.) Cass. inoculated with vesicular-arbuscular mycorrhizal fungi. New Phytol. 95:405-417.

69. Saif, S. R. 1986. Vesicular-arbuscular mycorrhizae in tropical forage species as influenced by season, soil texture, fertilizers, host species and ecotypes. Angew. Botanik 60:125-139.

70. Schenck, N. C., and Kinloch, R. A. 1980. Incidence of mycorrhizal fungi on six field crops in monoculture on a newly cleared woodland site. Mycologia 72:229-443.

71. Schenck, N. C., Siqueira, J. O., and Oliveira, E. 1989. Changes in the incidence of VA mycorrhizal fungi with changes in ecosystems. Pages 125-129 in: Interrelationships between microorganisms and plants in soil. V. Vancura and F. Kunc, eds., Elsevier, New York, NY.

72. Schwab, S. M., Johnson, E. L. V. , and Menge, J. A. 1982. Influence of simazine on formation of vesicular-arbuscular mycorrhizae in *Chenopodium quinona* Wild. Plant Soil 64:283-294.

73. Schwinn, F. J., and Staub, T. 1987. Phenylamides and other fungicides against Oomycetes. Pages 259-273 in: Modern Selective Fungicides-Properties, Applications, Mechanisms of Action. H. Lyr. Longman, ed., Scientific Technical. 383 pp.

74. Sieverding, E. 1991. Vesicular-Arbuscular Mycorrhiza Management in Tropical Agrosystems. Deutsche Gesellschaft fur Technische Zusammenarbeit. Eschborn. 371 pp.

75. Sieverding, E., and Leihner, D. 1984. Effect of herbicides on population dynamics of VA-Mycorrhiza with cassava. Angew Botanik 58:283-294.

76. Spokes, J. R., MacDonald, R. M. and Hayman, D. S. 1981. Effects of plant protection chemicals on vesicular-arbuscular mycorrhizae. Pestic. Sci. 12:346-350.

77. Sreenivasa, M. N., and Bagyaraj, D. J. 1989. Use of pesticides for mass production of vesicular-arbuscular mycorrhizal inoculum. Plant Soil

119:127-132.

78. Struble, J. E., and Skipper, H. D. 1988. Vesicular-arbuscular mycorrhizal fungal spore production as influenced by plant species. Plant Soil 109:277-280.

79. Strzemska, J. 1975. Mycorrhiza in farm crops grown in monoculture. Pages 527-536 in: Endomycorrhizas. F. E. Sanders, B. Mosse, and P. B. Tinker, eds., Academic Press. New York.

80. Sutton, J. C., and Sheppard, B. R. 1976. Aggregation of sand-dune soil by endomycorrhizal fungi. Can. J. Bot. 54:326-333.

81. Sylvia, D. M., and Neal, L. H. 1990. Nitrogen affects the phosphorus response of VA mycorrhiza. New Phytol. 115:303-310.

82. Thompson, J. P. 1987. Decline of vesicular-arbuscular mycorrhizae in long fallow disorder of field crops and its expression in phosphorus deficiency of sunflower. Aust. J. Agric. Res. 38:847-867.

83. Thomson, B. D., Robson, A. D., and Abbott, L. K. 1986. Effects of phosphorus on the formation of mycorrhizas by *Gigaspora calospora* and *Glomus fasciculatum* in relation to root carbohydrates. New Phytol. 103:751-765.

84. Tisdal, J. M. 1991. Fungal hyphae and structural stability of soil. Aust. J. Soil Res. 29:729-743.

85. Toth, R., Page, T., and Castleberry, R. 1984. Differences in mycorrhizal colonization of maize selections for high and low ear leaf phosphorus. Crop Sci. 24:994-996.

86. Wallace, L. L. 1987. Effects of clipping and soil compaction on growth, morphology and mycorrhizal colonization of *Schizachyrium scoparium*, a C4 bunchgrass. Oecologia 72:423-428.

87. Wright, S. F., and Morton, J. B. 1989. Detection of vesicular-arbuscular mycorrhizal fungus colonization of roots by using a dot-immunoblot assay. Appl. Environ. Microbiol. 55:761-763.

88. Vivekanandan, M., and Fixen, P. E. 1991. Cropping systems effects on mycorrhizal colonization, early growth, and phosphorus uptake of corn. Soil Sci. Soc. Am. J. 55:136-140.

89. Vilarino, A., and Arines, J. 1992. The influence
    of aqueous extracts of burnt or heated soil on the
    activity of vesicular-arbuscular mycorrhizal fungi
    propagules. Mycorrhiza 1:79-82.
90. Yocom, D. H., Larsen, H. J., and Boosalis, M. G.
    1985. The effects of tillage treatments and a
    fallow season on VA mycorrhizae of winter wheat.
    Page 297 in: Proc. Sixth North American Conf. on
    Mycorrhizae. R. Molina, ed., Bend, OR. 25-29 June
    1984. Forest Research Laboratory, Corvallis, OR.

EFFECTS OF NURSERY CULTURAL PRACTICES ON MANAGEMENT OF SPECIFIC ECTOMYCORRHIZAE ON BAREROOT TREE SEEDLINGS

Charles E. Cordell
USDA Forest Service
Asheville, North Carolina

and

Donald H. Marx
USDA Forest Service
Athens, Georgia

A variety of soil factors and nursery cultural practices affect ectomycorrhizal fungi and their tree seedling hosts. Soil factors include pH, drainage and moisture, texture, fertility, irrigation, organic matter, and microorganisms. A highly significant factor involves the basic characteristics of the ectomycorrhizal fungus/tree host symbiotic relationship. This includes proper fungus/host species selections; fungus/host developmental stage compatibility (early or late stage fungus); fungus isolate variability, source, and age; and fungus inoculum effectiveness. Nursery cultural practices involve all procedures utilized to produce the seedling hosts and include nursery and seedbed location selections, fungus and host compatibility selections, pH (soil and water) adjustments and maintenance, fertilization, irrigation, seedling root pruning, shading, and harvesting. Certain pesticides (primarily fungicides) have either positive or negative effects on ectomycorrhizal development and maintenance. A successful ectomycorrhizal nursery management program does not end with the shipping of seedlings from the nursery. Special consideration and effort are required for the

retention of the carefully nurtured nursery ectomycorrhizae throughout seedling handling, storage, shipping, and planting. Guidelines for successful nursery
management of ectomycorrhizae, therefore, are primarily aimed at developing and maintaining high-quality
host seedling root systems rather than meeting the
requirements of selected species of ectomycorrhizal
fungi.

There is no question that managing mycorrhizal
fungi as well as seedlings adds to the complexity of
the nurseryman's job. Instead of providing for the
needs of only the tree seedlings, he now has to provide for the seedlings and their symbiotic fungi. In
most instances, however, the adjustments that must be
made are not as large as might be expected. Ectomycorrhizal fungi have the same general moisture,
temperature, fertility, and pH requirements as their
tree hosts. In this chapter, the special procedures
and precautions that must be taken in order to artifi-
cally introduce specific species of ectomycorrhizal
fungi and maintain them on seedlings are described.
Also, bareroot nursery seedling production in the
Southeastern United States is emphasized where approximately 75% of United States forest tree seedlings are
produced. A comprehensive description of mycorrhizae
in relation to container seedling production is provided by Castellano and Molina (1). It all starts
with the ectomycorrhizal fungus selection. Both the
fungus species and the isolate must be chosen with
care. Pesticides must be used judiciously, and fertilization and cultural practices may require alteration to some extent. The results are well worth the
trouble, however. A special satisfaction is realized
by knowing that by expending some extra effort, seedlings can be produced with an improved chance of
survival and rapid growth after outplanting. The
production of seedlings that perform well in the field
is the primary objective of these customized nursery
practices.

FUNGUS CHARACTERISTICS

A primary consideration in any ectomycorrhizal
fungus nursery management program is the selection of

a fungus species. If all fungus species were equally beneficial, there would be little point in comparing or discussing them. Physiological and ecological differences among fungus species are criteria for selecting one. Some species are early-stage fungi and are primary symbionts on tree seedlings. Others are late-stage and are primary symbionts on older trees. Fungi for nursery applications should be early-stage with the physiological capacity to form abundant ectomycorrhizae on seedlings of the desired hosts. Isolates of a given species also vary considerably, so the suitability of the isolate for the tree species and planting situation should be verified. Fungus isolates that are ill-adapted, either ecologically or to the host species, should be viewed with suspicion. Isolates that have been grown in pure culture for extended periods also should be avoided (12). The selected fungus must also have the capability to grow rapidly in pure culture and withstand substantial physical, chemical, and biological manipulation in the nursery (17). The fungus inoculum must also be able to survive in either bareroot nursery soil or container growing medium for a minimum of 4 to 6 weeks between fungus inoculation and the subsequent production of feeder roots by the developing seed-lings. A desirable fungus also should have the capacity to survive several weeks of storage between fungus inoculum production and use.

An important limiting factor is the ecological adaptation of the selected fungus to the type of site on which the host seedlings are scheduled to be plant-ed. The ecological adaptability of an ectomycorrhizal fungus is closely correlated with the metabolic path-ways it has evolved to contend with environmental variation (24). The selected fungus must be able to adapt to a wide variety of environmental conditions, including extremes of soil and climate, antagonism from other soil organisms (including other ecto-mycorrhizal fungi), pesticides, physical disruption of fungus mycelium by nursery culture practices, and the abrupt and severe physiological adjustment from an intensively cultured and managed nursery soil to an alien field site with all its natural stresses.

Another desirable fungus characteristic is the production of hyphal strands or rhizomorphs to enhance nutrient and water absorption during nutrient or moisture stress periods on the field sites. Of course, the selected fungus must be aggressive in its growth habits and be capable of forming abundant ectomycorrhizae on host seedlings immediately after feeder roots are produced.

Marx and others (16) found a positive correlation between inoculum application rates and types and subsequent ectomycorrhizal development on loblolly pine (*Pinus taeda* L.) seedlings by *Pisolithus tinctorius* (Pers.) Coker & Couch (Pt). Higher rates of vegetative inoculum produced increased amounts of ectomycorrhizae and the vegetative inoculum produced more ectomycorrhizae than spore inoculum (Table 1). Studies also have repeatedly shown that Pt ectomycorrhizae must at least equal the total of all other ectomycorrhizae on the seedling feeder roots (Pt index = 50) in order to significantly improve tree survival and their early growth in field plantings (16). Finally, the selected fungus should be capable of maintaining superiority over the naturally occurring fungi on the seedling roots--both in the nursery and, at least, during the early establishment and developing years of the field plantings (5).

SOIL OR GROWING MEDIUM

The composition of the nursery soil or the container growing medium can significantly affect the development of ectomycorrhizae on host tree seedlings. Like their seedling hosts, most ectomycorrhizal fungi are favored by the coarse-textured sandy or sandy loam nursery soils and container growing media with adequate composition to promote good internal water drainage and aeration. Growing media for container seedlings vary widely (23) and those with a high proportion of coarse-textured particles (vermiculite, pine bark, etc.) promote ectomycorrhizal development (21).

The pH of the nursery soil, container growing medium, and irrigation water is important for ecto-

Table 1. Effect of Pt inoculum type and rate on production of Pt ectomycorrhizae on loblolly pine seedlings in a South Carolina nursery.

Inoculum treatment	% Ectomycorrhizae Pt	All fungi	% Seedlings with Pt	Pt index[y]	Number Pt fruit bodies/m^2
IMRD:0.33 L/m^2	56a[z]	65a	100a	85a	3.2a
IMRD:0.16 L/m^2	24bc	56b	58cd	27cd	1.4c
Spore encap. seeds	39b	58b	92b	62b	2.8ab
Spore pellets	15cc	52b	44c	13e	0.6c
Spores sprayed	26bc	53b	66cd	33c	1.4c
Control	<1d	49c	2d	<1f	0

[y]Pt index = a(b/c) where:
  a = % of seedlings with Pt ectomycorrhizae;
  b = Average % of feeder roots with Pt ectomycorrhizae;
  c = Average % of feeder roots with ectomycorrhizae formed by Pt and
      other fungi.
[z]Means followed by same letter within columns are not significantly different at $\underline{P}$ = 0.05.

mycorrhizal development. Most fungi have pH requirements similar to their hosts, but some can tolerate
unusual deviations (acid strains of Pt can tolerate pH
3). Frequently, acidic or alkaline amendments are
needed to meet the pH requirements of the selected
ectomycorrhizal fungus, as well as the host seedlings.

High soil organic matter promotes internal soil
water drainage, aeration, and seedling root development. All these factors improve ectomycorrhizal
development and longevity because roots are precursors
for ectomycorrhizal development. Soil organic matter
can be increased with cover crops and mulching in the
nursery. Precautions are needed with mulching materials. Some have adverse effects on ectomycorrhizal
fungi or host seedlings. Fresh sawdust can radically
change soil carbon:nitrogen ratios, and with high
populations of bacteria, soil nitrogen can become
deficient. Animal manure and sewage sludge may
contain components (high ammonia and heavy metals)
that are toxic to tree seedlings or their
ectomycorrhizal symbionts.

High fertilization rates and high lime amendments
required to increase the soil fertility and pH for

cover crops such as corn, soybeans, and sorghum-sudan may have high residual pH or fertility effects on the subsequent seedling crop. Excessively high pH or fertility can, as previously described, have significant negative effects on ectomycorrhizal development and seedling quality.

Artificially introduced ectomycorrhizal fungi generally respond quite differently in artificial container media than in nursery soils. With reasonable precautions, development of ectomycorrhizae has been excellent on container seedlings. Conversely, the best natural ectomycorrhizal development has been obtained in nursery soils. Inoculation is successful in container media because competing natural-origin ectomycorrhizal fungi, along with many other microorganisms, are absent. Nursery soils typically have a variety of microorganisms including bacteria, fungus parasites, fungus saprophytes, and a natural component of native ectomycorrhizal fungi. All too frequently, these resident soil microorganisms successfully compete with the introduced ectomycorrhizal fungi. Consequently, soil sterilization or fumigation is needed for consistent success in the introduction of ectomycorrhizal fungi into nursery soils (19).

NURSERY CULTURAL PRACTICES

The quantity and quality of ectomycorrhizae in bareroot and container nurseries can be significantly increased through nursery cultural practices. Guidelines for ectomycorrhizae management are primarily aimed at developing and maintaining high-quality seedling root systems. Enhanced quality and quantity of ectomycorrhizal fungi are closely correlated with increased seedling quality. Ectomycorrhizal fungi generally respond to the same cultural practices as their tree seedlings. Therefore, with few exceptions, any cultural practice that helps host seedlings is also likely to help its ectomycorrhizae (5).

Acidity is particularly important in the production and maintenance of ectomycorrhizae. Recent studies (11) show the effects of soil pH (pH 4.8, 5.8, and 6.8) on Pt ectomycorrhizal development of loblolly pine (Figure 1). Vegetative inoculum of Pt buried in nursery soil lost significant viability after 54 days exposure to soil pH 6.8. After one growing season in the nursery, Pt development on 1-0 loblolly pine seedlings in soil at pH 6.8 was only about 25 percent of that in more acid treatments. An alkaline nursery soil or container growing medium severely restricts development of certain ectomycorrhizal fungi. Coincidentally, high soil or growing medium pH (above 6.0) may significantly reduce seedling root growth and subsequent ectomycorrhizal development. In addition, pH levels above 5.5 increase the severity and damage potential of soil-borne, root pathogenic fungi (4). Desired pH levels are routinely maintained in nurseries by applying elemental sulfur or ammonium sulfate to reduce the pH or lime to increase it.

The pH of irrigation water can be reduced (high water pH levels are most frequently encountered) by metering prescribed amounts of phosphoric or sulfuric acid into the irrigation water (4). Water above pH 6.5 and water containing adverse agents (minerals, pathogenic fungal spores, nematodes, and weed seeds) must be modified. Also, frequent or excessive irrigation, particularly in containers and poorly-drained nursery beds, may have significant negative effects on soil aeration, thus impacting root and ectomycorrhizal development.

The type, rate, and timing of fertilizer applications can significantly affect tree seedlings and their ectomycorrhizal symbionts. Two primary minerals and major fertilizer nutrient components that have received considerable attention are nitrogen and phosphorus. Marx (11) studied the effects of nitrogen fertilization (three, six, or nine applications of 50 kg N/ha as NH_4NO_3) on Pt ectomycorrhizal development of loblolly pine (Figure 1). The highest applications of NH_4NO_3 (450 Kg N/ha) resulted in nursery soil concentrations of NO_3-N ranging from 60-120 kg/ha and of NH_4-N ranging from 90-130 kg/ha at pH 4.8 and 5.8 and provided the highest amounts of Pt ectomycorrhizal development (Figure 1). Also, increased nitrogen applications resulted in corresponding increases in Pt development on 1-0 loblolly pine seedlings at the

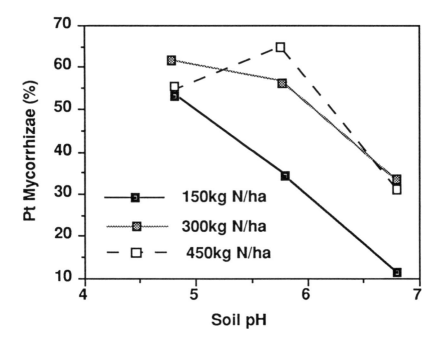

Figure 1. Effects of pH and nitrogen fertilization on Pt ectomycorrhizal
development on 1-0 bareroot loblolly pine seedlings.

higher pH levels (5.8 and 6.8) relative to the significantly reduced Pt development obtained on seedlings
receiving the lowest (150 Kg N/ha) nitrogen rate (11).
Higher than normal phosphorus levels (above 170 Kg/ha
available P--Bray II analysis) have been suggested to
inhibit ectomycorrhizal development in nurseries (4).
However, Marx's results (11) and those of numerous
fungus inoculation studies (17,18) failed to show any
distinct negative effects of highly variable levels of
nitrogen, phosphorus, and several other associated
macro- and micro-fertilizer components on Pt
ectomycorrhizal development. The studies involved
over 20 species of conifer seedlings in over 40 nurseries across the United States and in Canada. Negative effects of high levels of nitrogen or phosphorus
have been associated with increased pH. Molina and

Chamard (20) observed that different slow-release
fertilizer treatments did not significantly alter
ectomycorrhizae formation by *Laccaria laccata* (Scop.
ex Fr.) Cook on container-grown Douglas-fir
Pseudotsuga menziessi (Mirb.) Franco seedlings and
that Ponderosa pine (*Pinus ponderosa* Dougl. ex Laws.)
container seedlings had 95 percent ectomycorrhizal
short roots over all fertility treatments. In contrast, Castellano et al. (2) observed that a
slow-release fertilizer suppressed ectomycorrhizal
formation by *Rhizopogon vinicolor* Smith and *R. colos-*
sus Smith on container-grown Douglas-fir, but abundant
similar ectomycorrhizae developed with both high and
low levels of soluble fertilizers. Ruehle and Wells
(22) developed a custom formulation of soluble
fertilizer that provided highly consistent Pt
ectomycorrhizal development on container-grown
loblolly pine seedlings.

Seedling shading can significantly reduce ectomycorrhizal development and seedling quality. The
photosynthetic potential of the seedling influences
the susceptibility of tree roots to ectomycorrhizal
development. High light intensity promotes ectomycorrhizal development, and light intensity below 20% of
full sunlight significantly reduces ectomycorrhizal
development (10). Photosynthates supplied to the
fungal symbiont by the tree host are of paramount
importance to the development, function, and maintenance of ectomycorrhizae. These carbon sources fur-
nish the energy for root and fungal growth and are
closely associated with nutrient uptake by the ectomycorrhizal feeder roots (7).

Lateral and vertical root pruning improve the
development of feeder roots and their associated ectomycorrhizae by increasing root fibrosity (8) (Figure
2). Root pruning produces wounds that result in the
redirection of primary carbon supplies from the
seedling foliage to the roots. The increased energy
supply stimulates development of feeder roots and
associated ectomycorrhizae (Figure 2).

Hatchell (8) observed significant increases in
feeder roots and Pt ectomycorrhizae on 1-0 bareroot
longleaf pine (*Pinus palustris* Mill.) after lateral

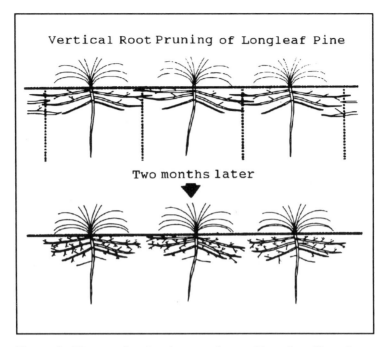

Figure. 2. Diagram showing the procedure and benefits of lateral root pruning 1-0 bareroot longleaf pine seedlings in southern forest tree nurseries.

and horizontal root pruning. In addition, a combination of root pruning, reduced seedbed density, and inoculation with Pt in the nursery resulted in significant increases in field survival and height growth of longleaf pine during the first 4 years in the field (8) (Table 2).

PESTICIDES

Chemical soil fumigants are nonselective biocides. They kill a variety of soil organisms, including pathogenic, saprophytic, and mycorrhizal fungi; insects; nematodes; and weed seeds. The most commonly used fumigant in forest nurseries in the USA is the 98% methyl bromide, 2% chloropicrin mixture. This fumigant does an excellent job of preparing the nursery soil for introduction of specific ectomycorrhizal

Table 2. Effects of selected nursery cultural practices on longleaf pine field survival and growth after 4 years.

Treatment		Survival (%)	Height (ft)	Relative Plot Vol.
No lateral root pruning No Pt ectomycorrhizae	12 Seedlings/ft^2	53	2.4	1.3
	15 Seedlings/ft^2	40	2.4	1.0
Lateral root pruning Pt ectomycorrhizae	12 Seedlings/ft^2	83	2.8	2.5
	15 Seedlings/ft^2	81	2.5	2.3

fungi (19). A 67% methyl bromide 33% chloropicrin formulation (MC-33) has been particularly effective in controlling the more persistent soil-borne root pathogenic fungi (3). Effective soil fumigation either eliminates or significantly reduces all living organisms in the nursery soil or container growing media in the zone of effective fumigant concentration. The beneficial ectomycorrhizal and saprophytic fungi are considered pioneer organisms in the fungal succession and invasion of these highly disturbed soils. Consequently, these fungi usually rapidly invade recently treated nursery soils and container growing media where adjacent fungus inoculum sources are available. Frequently, ectomycorrhizal fungi build up to higher population densities than those found in untreated soil or growing medium (3). To be most effective, soil fumigation (spring fumigation for spring seed sowing) should coincide as much as possible to scheduled seed sowing. Spring soil fumigation maximizes the reduction of natural-origin ectomycorrhizal fungi (i.e., *Thelephora terrestris* (Ehrh.) Fr.) with their optimum spore production and dispersal periods in the spring. Spring fumigation also minimizes the potential for seedbed contamination when closely followed with seed sowing. Therefore, an effective spring soil or growing media fumigation procedure coordinated with an effective nursery ectomycorrhizae (inoculated or natural-origin) management program will significantly increase the quality and quantity of selected ectomycorrhizae on the seedling hosts. Contrary to previous myths concerning the negative effects of soil fumigation on ectomycorrhizal development, more recent results consistently show the rapid development (6 to 8 weeks) from adjacent natural spore sources on feeder roots following fumigation (17).

Certain fungicides affect ectomycorrhizal development and maintenance in nurseries (25). Triadimefon (Bayleton), a systemic fungicide used to control the fusiform rust fungus (*Cronartium quercuum* (Berk.) Miyabe ex Shirai f. sp. *fusiforme* Burds and Snow) in southern pine nurseries, significantly suppresses Pt and natural ectomycorrhizal development, even at dosages below those recommended for rust control (Figure 3) (13,15). The negative effects of Bayleton on ectomycorrhizae are of two primary types. The first type results from negative effects of Bayleton sprays translocated from seedling foliage to roots either as Bayleton or its by-product Baytan. Second, is negative effects of Bayleton leaching down through soil and inhibiting inoculum sources (spores or vegetative mycelium). Bayleton significantly reduced the viability of vegetative inoculum of Pt buried in nursery soil (13). Triadimefon sprays on established seedlings also reduced Pt and natural ectomycorrhizae, fungus fruit body production, and pine seedling fresh weights (Table 3). A Bayleton seed soak, however, did not have negative effects on ectomycorrhizae, fruit body production, or seedling size (13). Two other fungicides, Benomyl and Captan, stimulate Pt and natural ectomycorrhizal development and maintenance in Southern U.S. pine nurseries (19,25). Consequently, these latter two beneficial fungicides are recommended as root disease control agents and as stimulators of ectomycorrhizal development on seedlings following inoculation of seedbeds with selected ectomycorrhizal fungi.

Very little information is available about the effects of insecticides and herbicides on

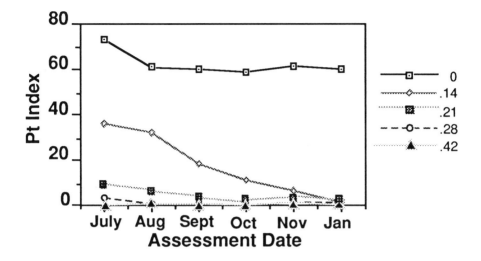

Figure 3.   **Effect of triadimefon (Bayleton) dosage rates on Pt ectomycorrhizae on 1-0 loblolly pine seedlings in a South Carolina nursery. Triadimefon dosage rate = kg a.i./ha.**

Table 3. Effects of two fungicides on ectomycorrhizal development, fungus fruiting body production, and loblolly pine seedling fresh weight in a South Carolina nursery.

	% Ectomycorrhizae		Seedling fresh	Ectomycorr. fruit bodies	
	Pt	All fungi	weights (g)	Pt	Other
Bayleton[R] + P. tinctorius	0.7	15.3	11.1	57	208
Fermate + P. tinctorius	34.7	46.3	13.4	442	497
Check (Uninoculated)					
Bayleton[R]	0.1	17.7	10.7	3	322
Fermate	1.5	41.7	12.3	10	640

ectomycorrhizae. Present research results are quite variable and conflicting. Herbicides may affect the host seedling by damaging foliage or roots and indirectly reducing ectomycorrhizal development (25).

SEEDLING HARVESTING AND HANDLING

It is not enough to stimulate ectomycorrhizal development. Once mycorrhizae have developed, detailed procedures are needed to maintain them in the nursery and during transport to the field. Special care must be taken during all stages of seedling harvesting and handling to maintain maximum amounts of feeder roots and ectomycorrhizae. Ectomycorrhizae are delicate structures that can be rather easily ripped off during harvesting (lifting), desiccated in storage, or cut off prior to planting. To preserve seedling quality, harvesting and handling techniques must be modified to minimize damage to feeder roots and ectomycorrhizae (5). Marx and Hatchell (14) studied the effects of various degrees (0, 33, 66, and 100%) of feeder root and ectomycorrhizal removal on 1-0 loblolly pines. Severe root stripping (66% removal) significantly reduced development of new lateral roots and ectomycorrhizae with subsequent severe tree mortality in a field planting (Table 4) (14).

Whenever seedlings are being handled, special care is required to avoid drying of the feeder roots and delicate ectomycorrhizae by exposure to wind and sun. Numerous studies have documented the effects of various seedling storage and shipping practices (packing materials, storage time, storage and shipping temperatures) (9). Improper transportation to the planting site or rough handling during planting also can damage feeder roots and ectomycorrhizae, reducing seedling quality and subsequent field survival. Tree planters must be made aware of proper handling and planting procedures and the reasons for such measures. Where possible, seedlings should be continuously refrigerated from nursery storage to field planting. With either machine or hand planting, root pruning at the planting site should be avoided. This practice

Table 4. Effects of root stripping of ectomycorrhizae on lateral root and ectomycorrhizal development; field survival and growth of loblly pine seedlings.

Ectomycorrhizal condition	Removal (%)	Lateral roots	Pt	All fungi	Survival (%)	Volume (cm)[y]
Pt	0	5.2a[z]	22a	34a	97a	490a
Ni		2.5b	0	29ab	93a	364c
Pt	33	4.0a	18a	23b	94a	451a
Ni		1.3b	0	21b	94a	312x
Pt	66	3.6ab	6b	13c	93a	324ab
Ni		0.3c	0	10c	85b	282cd
Pt	0	8.6a	34a	41a	95a	
Ni		6.0ab	0	44a	91a	
Pt	33	9.8a	28ab	36a	93a	
Ni		4.0c	0	38a	93a	
Pt	66	7.8ab	9b	17b	90a	
Ni		2.8c	0	17b	82b	
Pt	0	16.6a	39a	53a	76a	
Ni		15.7a	0	39ab	68a	
Pt	33	17.8a	31a	48ab	68a	
Ni		9.0b	0	51a	55ab	
Pt	66	13.8ab	12b	44ab	65a	
Ni		7.8b	0	31b	52ab	

Pt = Pisolithus tinctorius; NI = naturally inoculated in nursery.
[y]Volume based on tree measurements after 2 years in field.
[z]Means followed by the same letter within columns are not significantly different at $\underline{P}$ = 0.05.

removes significant amounts of carefully nurtured host seedling feeder roots and the associated ectomycorrhizae. Even a few minutes of direct exposure to high temperatures, high winds, and low humidity can injure feeder roots and ectomycorrhizae. A top priority in planting should be to preserve seedling quality and vigor by maintaining feeder roots and ectomycorrhizae (5,6,9).

CONCLUSIONS

Ectomycorrhizal fungi are generally affected by the same conditions and practices that affect the host tree seedling. As a result, an ectomycorrhizae management program requires only minor modifications to those previously established for seedling cultural practices. Selection of the proper ectomycorrhizal fungus for the tree species and planting site is very important. The soil should be fumigated prior to inoculation. Soil pH cannot be permitted to rise above 6.0. Effects on ectomycorrhizal fungi should be considered when pesticides are chosen to control diseases or for other purposes. Even with the variety and complexity of factors involved, a little extra effort can produce consistently good results. Program success requires the cooperation of all concerned parties--nursery personnel, field foresters, land managers, and tree planters. Proper awareness and due consideration and care must be afforded the two players in this symbiotic association--the tree seedling and its complement of ectomycorrhizal fungi.

LITERATURE CITED

1. Castellano, M. A., Molina, R. 1989. Mycorrhizae. Pages 101-167 in: T. D. Landis, R. W. Tinus, S. E. McDonald, J. P. Barnett, eds. The Container Tree Nursery Manual, Vol. 5. USDA For Serv. Agric. Handb. 674. 171 pp.
2. Castellano, M. A., Trappe, J. M. and Molina, R. 1985. Inoculation of container grown Douglas-fir seedlings with basidiospores of *Rhizopogon vinicolor* and *R. colossus*: effects of fertility and spore application rate. Can J. For. Res. 15: 10-13.
3. Cordell, C. E. 1983. Effective soil fumigation. Pages 196-201 in: Proceedings: 1982 Southern Nursery Conferences. USDA For. Serv., State and Private Forestry, Region 8, Tech. Pub. R8-TP4. 312 pp.
4. Cordell, C. E., Anderson, R. L., Hoffard, W. H., Landis, T. D., Smith, R. S. Jr., and Toko, H. V. 1989. Forest Nursery Pests. USDA For. Serv. Agric. Handb. 680. 184 pp.
5. Cordell, C. E., Omdal, D. W., and Marx, D. H. 1990. Identification, management, and application of ectomycorrhizal fungi in forest tree nurseries. Pages 143-155 in: Proc. first meeting of IUFRO Working Party S2.07-09. Diseases and in-

sects in forest nurseries. Victoria, B.C. Canada.
298 pp.

6. Cordell, C. E., Owen, J. H., Marx, D. H. 1987.
    Mycorrhizae nursery management for improved seed-
    ling quality and field performance. Pages 105-115
    in: Meeting the challenge of the nineties:
    Proceedings of a meeting of the Intermountain
    Forest Nursery Association. USDA For. Serv. Gen.
    Tech. Rep. RM-151. 138 pp.

7. France, R. C. and Reid, C. P. P. 1983.
    Interactions of nitrogen and carbon in the physi-
    ology of ectomycorrhizae. Can. J. Bot. 61:
    964-984.

8. Hatchell, G. E. 1985. Nursery cultural practices
    affect field performance of longleaf pine. Pages
    148-156 in: Proc. Inter. Symp. of Nursery Manage.
    Practices for the Southern Pines. D. B. South,
    ed., Co-sponsored by School of Forestry, Ala.
    Agric. Exp. Sta., Auburn Univ. Alabama, and IUFRO
    Subject Group S3.202-03 Nursery Operations.

9. Lantz, C. W., Coord. 1989. A guide to the care
    and planting of southern pine seedlings. USDA
    For. Serv., Southern Region. Management Bull. R-
    8MB39. 44 pp.

10. Marx, D. H. 1991. The practical significance of
    ectomycorrhizae in forest establishment. Pages
    54-90 in: Ecophysiology of ectomycorrhizae of
    forest trees. The Marcus Wallenberg Foundation
    Symposia Proceedings: 7. The Marcus Wallenberg
    Foundation. S-791 80, Sweden. 90 pp.

11. Marx, D. H. 1990. Soil pH and nitrogen influence
    Pisolithus ectomycorrhizal development and growth
    of loblolly pine seedlings. For. Sci. 36:224-245.

12. Marx, D. H. 1981. Variability in ectomycorrhizal
    development and growth among isolates of *Piso-
    lithus tinctorius* as affected by source, age, and
    reisolation. Can. J. For. Res. 11:168-174.

13. Marx, D. H., and Cordell, C. E. 1987. Triadimefon
    affects *Pisolithus* ectomycorrhizal development,
    fusiform rust, and growth of loblolly and
    slash pines in nurseries. USDA For. Serv. Res.
    Pap. SE-267. 14 pp.

14. Marx, D. H., and Hatchell, G. E. 1986. Root

stripping of ectomycorrhizae decreases field performance of loblolly and longleaf pine seedlings. Southern J. Appl. For. 10:173-179.

15. Marx, D. H., Cordell, C. E., and France, R. C. 1986. Effects of triadimefon on growth and ectomycorrhizal development of loblolly and slash pines in nurseries. Phytopathology 76:824-831.

16. Marx, D. H., Maul, S. B., and Cordell, C. E. 1992. Application of specific ectomycorrhizal fungi in world forestry. Pages 78-98 in: Frontiers in Industrial Mycology. G. F. Leatham, ed., Chapman and Hall, New York. 222 pp.

17. Marx, D. H., Cordell, C. E., Kenney, D. S., Mexal J. G., Artman, J. D., Riffle, J. W., and Molina, R. J. 1984. Commercial vegetative inoculum of *Pisolithus tinctorius* and inoculation techniques for development of ectomycorrhizae on bare-root tree seedling. For. Sci. Monog. 25. 101 pp.

18. Marx, D. H., Ruehle, J. L., Kenney, D. S., Cordell, C. E., Riffle, J. W., Molina, R., Pawuk, W. A, Navratial, S., Tinus, R. W. and Goodwin, O. C. 1981. Commercial vegetative inoculum of *Pisolithus tinctorius* and inoculation techniques for development of ectomycorrhizae on container grown seedlings. For. Sci. 28(2):373-400.

19. Marx, D. H., Ruehle, J. L., and Cordell, C. E. 1991. Methods for studying nursery and field response of trees to specific ectomycorrhizae. Pages 383-411 in: Methods in Microbiology. Vol. 23. J.R. Norris, D.J. Read, and A.K. Varma, eds. Academic Press, London.

20. Molina, R., and Chamard J. 1983. Use of the ectomycorrhizal fungus *Laccaria laccata* in forestry. II. Effects of fertilizer forms and levels on ectomycorrhizal development and growth of container-grown Douglas-fir and ponderosa pine seedlings. Can. J. For. Res. 12:89-95.

21. Ruehle, J. L., and Marx, D. H. 1977. Developing ectomycorrhizae on containerized pine seedlings. USDA For. Serv. Res. Note SE-242. 8 pp.

22. Ruehle J. L., and Wells, C. G. 1984. Development of *Pisolithus tinctorius* ectomycorrhizae on container-grown seedlings as affected by

fertility. For. Sci. 30:1010-1016.
23. Tinus, R. W., and McDonald, S. E. 1979. How to grow tree seedlings in containers in greenhouses. USDA For. Serv. Gen. Tech. Rep. RM-60. 256.
24. Trappe, J. M. 1977. Selection of fungi for ectomycorrhizal inoculation in nurseries. Annu. Rev. Phytopathol. 15:203-222.
25. Trappe, J. M., Molina, R., and Castellano, M.A. 1984. Reactions of mycorrhizal fungi and mycorrhiza formation to pesticides. Annu. Rev. Phytopathol. 22: 331-359.

AIR POLLUTION AND ECOSYSTEM HEALTH: THE MYCORRHIZAL CONNECTION

Steven R. Shafer
North Carolina State University, ARS-USDA
Raleigh, North Carlinia

and

Michele M. Schoeneberger
University of Nebraska, Forest Service - USDA
Lincoln, Nebraska

Man's activities have influenced the earth's atmosphere since primitive humans intentionally maintained fires. In the past two centuries, however, processes associated with the Industrial Revolution and modern intensive farming have accelerated those changes. Changes in the chemical characteristics of the atmosphere as a result of release of gases, aerosols, and particulates into the air have been known for centuries. Recently, however, attention has turned to the physical changes that also may be occurring. It is unrealistic to assume that the amounts and types of chemicals discharged by human activities into the atmosphere do not alter the physical characteristics to some extent. The nature and magnitude of the physical changes, however, is still widely debated.

Many responses of individual plants to various airborne chemicals, water availability, and temperature are fairly well-known phenomena. Clear documentation of responses of whole populations, communities, and ecosystems to stresses is not so easily obtained, due to the increasing complexity of the systems with increasing spatial scale. Interpreting responses of a single plant to a gaseous

pollutant such as ozone (O_3), a phytotoxic
photo-oxidant, is comparatively easy in a
controlled-environment chamber in a laboratory or
greenhouse, compared to interpreting responses of a
group of plants growing in soil in an open-top field
chamber, for instance. In the field chamber, many
environmental characteristics other than O_3 dose must
be considered, such as variations in soil fertility,
soil depth, distribution of moisture availability in
the plot, etc. Those sources of variation can be
controlled somewhat better in the laboratory than in
the field.

On the same or larger scale, interactions of
plants with other organisms may alter expected
responses of a particular plant or group of plants to
the stress under study. This is particularly
important for those interactions among organisms that
involve transfer of some resource from one organism to
the other. Perhaps the most sensitive of these
situations include obligate symbioses. Obligately
symbiotic relationships are finely tuned, and any
stress that affects one of the partners presumably
affects the other.

The success of the mutualism between the plant
and fungus in a mycorrhizal relationship hinges in
large part on carbon flow from plant to endophyte. A
disruption of that relationship has major consequences
for both organisms, and any disrupting stress on a
large spatial scale hypothetically has consequences
for the community and ecosystem. This may be
particularly true for the arbuscular and
vesicular-arbuscular mycorrhizal (sensu 67; hereafter
collectively called VA mycorrhizal or VAM)
associations, which involve fungi that are
ecologically obligate symbionts.

The chemicals commonly designated "air
pollutants" and the possible physical changes in the
earth's atmosphere that might accompany them are
collectively the agents and consequences of so-called
"global climate change". These changes in the
environment may alter the relationships between plants
and mycorrhizal fungi. The nature of the changes in
the mycorrhizal condition, consequences for each

endophyte, subsequent significance to ecosystem
structure and function, and the limitations of
research on these topics to date are the subjects for
this review.

The idea that air pollutants may alter
relationships between plants and mycorrhizal fungi is
not as arcane as one might suppose. This concept
recently appeared in the popular press (19) in a brief
article entitled "Fewer Fungi Bode Poorly for
Forests". The major points of that article were:

1.   Fewer fungal fruiting bodies are being found in
     European forests than in the past, and air
     pollution is a primary suspect in their decline.
2.   A decrease in fungal populations spells trouble
     for the trees because the trees are dependent
     upon the fungi for nutrients and water.
3.   Without the mycorrhizal fungi, trees may have
     increased susceptibility to other stresses.
4.   It is not clear whether the decline of the fungi
     is a cause or a consequence of forest decline.
5.   "When fungi begin disappearing, it's certain that
     trees are in trouble."

The inferences and conclusions in this article might
be debated among scientists who study mycorrhizae,
fungal ecology, and/or forest decline and tree health,
although these ideas have been stated formally in the
scientific literature (2). That the public is made
aware of mycorrhizae, their importance in ecosystems,
and their potential relevance to air pollution biology
is not so debatable.

Mycorrhizae are comprised of tissue from two
organisms, so studying the impact of any environmental
factor on the association becomes a puzzle of
distinguishing a plant response, a fungal response,
and a response peculiar to the association. The
technical constraints of air pollution studies (i.e.,
usually limitations in space and numbers of plants)
severely restricts sampling and hence, interpretation
of results (7). Conclusions also may be specific to a
particular plant-fungus association, and
interpretation may be biased by knowledge or ignorance
of many aspects of the plant-soil relationship.
Moreover, considerable genetically-controlled

variation occurs in plant sensitivity and response to pollutants; this variation further complicates interpretation. All these limitations must be kept in mind as this topic is explored.

A corn (*Zea mays* L.) plant in a corn field is part of the crop; when soybeans (*Glycine max* (L.) Merr. are planted in that field the next spring and a corn plant emerges from a seed that fell during the autumn harvest, then the corn plant is a weed. Just as weed scientists often define a weed as "a plant out of place," any definition of an air pollutant depends on the viewpoint of an individual person. Many solid, liquid, and gaseous substances that can be termed "pollutants" occur naturally in the atmosphere, and many are beneficial or even required for living organisms. Their status as "pollutants" depends on a judgement of dose (concentration X duration of exposure) above the expected natural occurrence of the substance, geographical location of exposure, and expected condition of the receptor (plant, animal, soil, etc.) in a naturally occurring concentration that might be estimated for the substance. Consideration of the interactions of all the substances that have been studied as pollutants with respect to plant health and the implications for host-endophyte relationships are well beyond the scope of this review. Because much research on air pollution impacts on mycorrhizae has focused on O_3 and acid deposition, these two types of pollution will receive emphasis here. In addition, the potential importance of increases in concentrations of carbon dioxide (CO_2) to ecosystems warrants some discussion.

Consideration of these three types of air pollution includes three possible modes of action on mycorrhizae: change in carbon availability to the endophyte (i.e., fine roots themselves as infection sites, or quantities and types of soluble carbohydrates within infected roots); alteration of the rhizosphere environment in terms of chemical and microbiological characteristics; and alteration of nutrient balances within the plant, which indirectly affect mycorrhizal function. Unfortunately, each mode creates feedbacks into the system, creating a dilemma to be faced in interpreting data.

EFFECTS OF MAJOR AIR POLLUTANTS ON PLANTS AND MYCORRHIZAE

Ozone

Ozone is a natural component of the troposphere (the lowest approximately 10 km of the earth's atmosphere). Without man's activities, the concentration of O_3 during daylight hours probably would be approximately 30-50 parts per billion (ppb; 1 ppb=1 nL/L of air) (49). Peak concentrations in areas in which impacts on plants have been studied, however, have approached as high as 600 ppb (17). Ozone is a secondary pollutant, i.e., a byproduct of chemical reactions involving primary pollutants (materials that are considered pollutants and are released from an identifiable source) in the atmosphere. Ozone is created by photochemical reactions involving O_2, nitrogen oxides, organic aerosols, and peroxides. The last three precursors occur both naturally and anthropogenically, but the burning of fossil fuels and nitrogen transformation processes (microbiological and chemical) in soils involving animal manures and agricultural fertilizers release these substances in quantities that result in concentrations of tropospheric O_3 above the theoretical background level.

Ozone, like the nitrogen-based precursors, is a factor in global warming (104). Ozone is extremely phytotoxic and is the air pollutant of greatest concern in the United States with respect to agriculture (33). Impacts on forest trees also occur (17,59,65,85) and have been the focus of some large research programs (79). The many physiological and biochemical impacts of O_3 on plants have been extensively researched in recent decades, and a comprehensive discussion is well beyond the scope of this paper. Major effects on plants relevant in the context of mycorrhizae will be described briefly, but review articles and books can provide detailed information and specific literature citations (45,48,64,99).

Ozone is a strong oxidant that reacts with many substances in living organisms. The gas enters plant leaves through the stomata. In aqueous solution, O_3 produces O_2, hydroxyl ions, free radicals, and peroxides. Fatty acids, NADPH, and amino acids may be oxidized. An initial site of injury may be the cell membrane as O_3 disrupts the lipid layer. Physiological and ultrastructural evidence illustrates injury to membranes in chloroplasts as well. Oxidation of aromatic rings and sulfhydryl groups disrupts the structure and function of enzymes such as glucosyltransferases, cellulase, and amylase or starch phosphorylase. Other aspects of carbon physiology, such as suppression of glycolysis and glycolipid synthesis and stimulation of the pentose phosphate pathway and phenol metabolism, have been described as responses to O_3.

Such effects of O_3 at the cellular and biochemical level result in alterations of many aspects of carbon fixation, use, and allocation. Ozone-induced suppression of photosynthesis has been demonstrated since the 1950s for numerous plant species. In an important field study with 18-yr-old ponderosa pine (*Pinus ponderosa* Dougl. ex Laws) saplings in the San Bernadino Mountains of California, trees were classified into O_3 sensitivity classes based on foliar injury due to chronic stress of ambient O_3 ranging as high as 600 ppb. Based on different age classes of needles on the different sensitivity types, needle abscission occurred when photosynthetic rate was suppressed to approximately 10% of maximum. Since suppression of photosynthesis exceeded suppression of stomatal conductance, loss of chloroplast function probably contributed to the impact on photosynthesis (17). A summary of data comparing effects of O_3 on seven different crop and tree species, based on a standardized dose-metric in terms of concentration X hours of exposure, indicated that plants with inherently high photosynthetic rates (clover, soybean, poplar, wheat) appeared to be sensitive, whereas those plants with inherently slower rates (white pine, red oak) were relatively insensitive by comparison (75). This relationship may result from the relative stomatal conductances of the different species; high stomatal conductance allows a high O_3 flux into the leaf. Variation in sensitivity at the sub-species level, however, clearly indicates that factors in addition to stomatal conductance condition plant sensitivity.

The foliar injury induced by O_3 not only may reduce photosynthetic area but also may increase respiration in association with repair mechanisms. However, O_3 concentrations that do not induce injury also may stimulate respiration.

The importance of O_3 on the carbon economy of plants does not end with effects on photosynthesis and respiration. In most species, including herbaceous dicotyledonous plants, grasses, and perennials, exposure to chronic O_3 results in greater suppression of root growth than shoot growth (16,47), particularly during vegetative growth and for plants such as clover that store carbohydrates in crowns and roots (16). Although soil absorbs some O_3 from the air (98), movement into the soil is probably restricted to the top 2 centimeters (9). Thus, effects on roots arise from changes in carbon allocation, and these changes are readily detectable. For example, short-term exposure of ladino clover (*Trifolium repens* L. cv 'Tillman') to O_3 (50, 100, or 150 ppb, 4 h/day for 6 days) resulted in changes of photosynthesis ($^{14}CO_2$ uptake 48 h after the last exposure) of +10%, -22% or -85% respectively, compared to plants exposed to O_3-free air. The proportion of photosynthate allocated to the root system increased with O_3 from 32% for 0 ppb to 52% for 100 ppb, but then dropped sharply to 21% for 150 ppb. However, the root/shoot ratio (based on dry weight) steadily declined with increasing O_3 (8). Similarly, more ^{14}C from $^{14}CO_2$ was retained in leaves of tomato (*Lycopersicon esculentum* Mill.) plants exposed to O_3 compared to plants exposed to ozone-free air (56). When loblolly pine (*Pinus taeda* L.) seedlings were exposed during another study to O_3 (120 ppb, 7 h/day, 5 days/wk for 12 wk), no visible foliar injury occurred. Compared to plants maintained in charcoal-filtered air, however, seedlings exposed to O_3 exhibited reductions in photosynthesis (16%), speed of phloem transport (11%), phloem photosynthate concentration (40%), and total carbon transport to roots (45%); photosynthate that was retained above ground increased (94).

Changes in gross allocation are indicative of qualitative changes in the root tissues as well. Exposure to O_3 suppresses concentrations of soluble carbohydrates, starch, lipids, proteins, and phenols in roots.

The relevance of O_3-altered carbon allocation to VAM symbiosis has been clearly demonstrated in several studies. Exposure of tomato plants to 0 or 300 ppb O_3 (one 3-h exposure/wk for 4 wk) showed that O_3 could retard infection by *Glomus fasciculatum* (Thaxter) Gerd. & Trappe emend. Walker & Koske and that any growth benefit potentially provided by the VAM condition was offset by O_3. With short (less than 12 h) photoperiods, the VAM condition suppressed plant growth, and O_3 had no effect (58). In further work with the same host-fungus combination (56), plants were exposed to 0 or 150 ppb O_3 (two 3-h exposures/wk) or 300 ppb (one exposure/wk) for 9 wk. Compared to the controls, the 150 and 300 ppb treatments suppressed VAM levels by 46 and 63%, respectively. Growth of VAM plants was suppressed by O_3, but O_3 had no effect on nonVAM plants. When the tomato/*G. fasciculatum* system was exposed twice to 300 ppb O_3 (3 h each exposure, 7 days apart), root exudates and extracts contained lower concentrations of soluble amino acids than did exudates and extracts from plants exposed to ozone-free air. However, O_3 did not significantly affect the amino acid content of root exudates of nonVAM plants, while the amino acid content in root extracts from ozone-stressed nonVAM plants increased. Ozone treatments suppressed the amount of reducing sugars in exudates and extracts of VAM and nonVAM plants (56). Effects of O_3 on VAM tomato plants varied with fungal species, but O_3 suppressed root infection by all (though not always statistically significant). Again, VAM plants exhibited greater sensitivity to O_3 than did nonVAM

plants (57). In a field study, container-grown VAM
(*Glomus geosporum* Walker) and nonVAM soybean plants
were exposed in open-top chambers to O_3 (seasonal
means of 25, 49 or 79 ppb) for 139 days. In this
case, the VAM plants were less sensitive to O_3 than
nonVAM plants. Colonization of roots by the fungus
was not affected by O_3, but effects on the fungus were
demonstrated in that *G. geosporum* produced 40% fewer
chlamydospores/g of root at 79 than at 25 ppb (10).
Thus, O_3 suppresses carbon allocation to roots, and
the dependency of the VAM endophyte on plant-supplied
carbon renders the fungus indirectly subject to
O_3-induced stress.
     The effect of O_3 on roots is not a trivial
matter. In fact, change in root growth rate may
precede any above-ground symptoms of O_3 stress (66).
Thus, changes in the mycorrhizal status of the plant
may be impacted early in the plant-pollution
interaction and may actually condition other plant
responses to the stress.
     Chemical and microbiological characteristics of
the rhizosphere are important to mycorrhizal
development and function (24,34,63). The possibility
for O_3 to affect mycorrhizae through alterations in
the rhizosphere environment has not been documented
directly. Under experimental conditions, however, O_3
does alter rhizosphere characteristics. Exposure of
VAM tomato plants to O_3 (300 ppb, 3 h on each of 2
days 1 wk apart) reduced the quantity of soluble amino
acids and sugars in root exudates (56). Exposure of
bean (*Phaseolus vulgaris* L.) plants or hybrid
sorghum-sudangrass (*Sorghum* Moench) seedlings to O_3
caused changes in the population densities of bacteria
and fungi in the rhizosphere, including populations of
bacteria that exhibit phosphatase activity (53,81).
The consequences of effects on the rhizosphere for VAM
infection or function are unknown.
     Ozone and plant nutrients may interact and affect
the relationship between the plant and fungus.
Application of soluble phosphorus fertilizer to tomato
plants increased plant sensitivity to O_3, possibly
because the increased phosphorus concentration in the

foliage reduced the quantity of soluble carbohydrates and permitted increased foliar injury (50). In principle, this suggests how plant, VAM fungus, and pollutant interactions may be closely linked through phosphorus supply and demand.

Effects of O_3 on ectomycorrhizae have been inconsistent among different experiments. Ectomycorrhizae of paper birch (*Betula papyrifera* Marsh.) and loblolly pine seedlings inoculated with *Pisolithus tinctorius* (Pers.) Coker & Couch were apparently unaffected by O_3 in controlled exposures (42,52). In other studies, however, white pine (*Pinus strobus* L.) and northern red oak (*Quercus rubra* L.) seedlings produced more short roots and higher percentages of short roots that were mycorrhizal after exposure to O_3 than after exposure to O_3-free air (76,77). Similarly, in another study on Douglas-fir (*Pseudotsuga menziesii* (Mirb.) Franco) seedlings (28), the frequency of ectomycorrhizae increased with exposure to O_3. Such increases seem to contradict what might be anticipated from our understanding of effects of O_3 on roots until the data are considered carefully. When frequency was expressed as a percentage of 100 root tips examined (28), the increase may have reflected a suppression of carbon to roots that in turn suppressed root tip development. If the plant was more sensitive in terms of root tip production than the fungi were in terms of infection abilities, an increase of frequency of mycorrhizal tips might occur. In other experiments, O_3 suppressed ectomycorrhizal development of white pine (76) and loblolly pine (61). In a few experiments in which dose-response relationships between O_3 and ectomycorrhizae have been defined, the relationships have been either quadratic (77) or linear (61). Clearly, our understanding of O_3 effects on ectomycorrhizal systems is very poor.

Acid Deposition

Research on acid deposition and its effects on ecosystems has been extensive during the last 20 years. A recent compilation of findings (38) reviews much of the literature, and only representative

citations are provided here. The nitrogen oxides
(NO_x) that are produced from combustion of fossil
fuels and agricultural sources, in addition to natural
sources, contribute to global warming and are involved
in O_3 production (43). Sulfur dioxide also
contributes to global warming (104). Some of these
gases are phytotoxic and suppress plant growth
(64,78). Oxides of sulfur and nitrogen in gas and
aerosol forms are also precursors of dry acid
deposition and, after reaction with water vapor, wet
acid deposition. To date, little evidence supports
the hypothesis that acid deposition directly affects
plant growth, development, and yield except in
experiments with recurring treatments that are
considerably greater than current precipitation
acidities or H^+ deposition rates. Under experimental
conditions, impacts on plants have included erosion of
the foliar cuticle (27), ion exchange between droplets
on leaves and the leaves themselves (26), and
extremely variable effects on crop yield (37).
Possible effects on forests after long-term deposition
to poorly buffered forest soils and onto long-lived
trees, however, have not been conclusively discounted
and may contribute to declines in forest health and
productivity (15,80,90,91). Furthermore, indirect
effects of acid deposition on plants via impacts on
plant-microbe and plant-insect interactions have been
studied insufficiently for generalizations to be
suggested. An extensive literature review (100)
summarizes the various effects of acid deposition on
soilborne microorganisms and microbially mediated
processes.

Acid deposition involves simultaneous deposition
of several substances, H^+, NO_3^-, SO_4^{-2}, and other ions,
that can have different effects on plants and/or
fungi. For this reason, perhaps, impacts of acid
deposition on mycorrhizae have been consistent only
with respect to inconsistency. An important component
of acid deposition is nitrogen, and increased nitrogen
availability often decreases sugar concentrations in
roots (73). Accordingly, the number of mycorrhizal
short roots per lateral root on white pine seedlings
decreased with increases in the acidity of

NO_3-containing simulated rain, and percent colonization decreased as the nitrogen concentration in the plants increased (97). These observations were consistent with an experiment with loblolly pine exposed to simulated rains at pH 5.6, 4.0, or 3.2; each increase in rain acidity decreased the percentage of short roots that were mycorrhizal, as well as the number of mycorrhizae per cm of lateral root. However, the most acidic treatment (pH 2.4), reversed this trend and stimulated both shoot growth and mycorrhizal infection; interactions of nitrogen with other components of acid deposition or benefits of nitrogen deposition to the plants may have introduced a conflicting mechanism (84).

Acid deposition unquestionably alters the rhizosphere under experimental conditions. Changes induced by acid deposition on population densities of bacteria and fungi in the rhizosphere (81), however, could be plant-mediated or soil-mediated. One experiment indicated that effects of acid deposition on rhizosphere ecology was mediated little or none by impacts on foliage; effects of simulated acidic rain on fungal and bacterial population densities in the rhizosphere of hybrid sorghum x sudangrass seedlings were identical when simulated rain was applied to soil only or both foliage and soil, but rain acidity had no effects on populations when it was applied to foliage only (82). In contrast, a foliage-mediated effect on rhizobium nodules on roots of bean plants was demonstrated (41). Simulated acidic rain suppressed aspects of the host-parasite interactions of *Phytophthora cinnamomi* Rand. on blue lupin and of *Heterodera glycines* Ichinohe, *Meloidogyne hapla* Chitwood, and *M. incognita* (Kafoid & White) Chitwood on soybean (83,86); pathogen reproduction was suppressed in all cases, but whether this was a consequence of direct impacts on the pathogens or was mediated by the host was not determined. Whereas impacts of O_3 on mycorrhizae are almost certainly plant-mediated, this generalization may not be true for acid deposition.

The multiple possibilities for impacts of acidic deposition on plants and associated microorganisms

probably contribute to the variety of effects reported
on mycorrhizae. In addition to those discussed above,
other results of controlled exposures have included:
an increase in the number of mycorrhizae on roots of
loblolly pine exposed to simulated acidic rain and
high soil Mg, accompanied by a change in the frequency
of occurrence of certain morphotypes (92); no
acid-deposition-induced change in the overall
frequency of infection by ectomycorrhizal fungi but a
shift in the occurrence of certain types (62); changes
in the frequency of certain morphotypes in association
with a reduction in fine root branching and external
hyphae (22); and suppression of formation of
ectomycorrhizae by *Pisolithus tinctorius* (Pers.) Coker
& Couch with white pine (96). The numerous possible
interactions among plant species and genotype,
endophyte, simulated rain formulation, deposition
rates and application frequencies, and soil chemical
and drainage characteristics are among the many
factors that probably contribute to inconsistencies in
experimental results.

Carbon Dioxide

Carbon dioxide (CO_2) is the gas that is of most
concern in considerations of global warming and
changes in precipitation amounts and distribution (43)
Primary anthropogenic sources are fossil fuel burning
and deforestation (including biomass burning and loss
of photosynthetic area). The estimated anthropogenic
release of CO_2 during 1860-1987 was 241 billion tons,
with 25% attributable to deforestation and other
changes in land use (104). In addition to the obvious
consequences for communities and ecosystems following
potential climate change (4), CO_2 has distinct
physiological consequences for individual plants.
Several detailed reviews are available (93,95).
Impacts on plants that could affect relationships with
mycorrhizal fungi include changes in photosynthesis,
biomass accumulation, and carbon allocation; water use
efficiency; and mineral nutrient demand. Alterations
in carbon allocation indicate changes in source-sink
relationships for carbon and suggest implications for
mycorrhizal fungi and other microorganisms associated

with roots. In a series of detailed analyses of root:shoot ratios, the response of two "root" crops radish (*Raphanus sativus* L.), carrot (*Daucus carota* L. var. *sativa* DC) to above-ambient concentrations of CO_2 differed from those of cotton (*Gossypium hirsutum* L.) and soybean; a near-doubling of CO_2 increased the root:shoot ratio of carrot and radish by an average of 36%, whereas root:shoot ratios for the two other species did not change significantly (36). In another study, however, soybean plants were exposed for an entire growing season to CO_2 concentrations ranging from 330 to 800 ppm; 3 mo after emergence, greater-than-ambient CO_2 had stimulated root biomass by 26-31% over that of plants grown in ambient CO_2, and cumulative root length was proportional to CO_2 concentration. Furthermore, CO_2 altered root architecture; the length of individual roots was not affected, but plants grown in the high concentrations of CO_2 had more root tips than those grown in ambient CO_2. Thus, CO_2 did not affect the volume of soil explored by roots, but did increase the intensity of root exploration within the volume (20). As carbon fixation, plant growth rate, and root characteristics change, dependence on mycorrhizal fungi for uptake of nutrients from soil may be altered substantially.

Few experiments have been conducted to examine the impact of increased atmospheric CO_2 on mycorrhizae. In one study, tulip poplar (*Liriodendron tulipifera* L.) seedlings that were infected by a VA mycorrhizal fungus were exposed to ambient (approximately 350 ppm) or 800 ppm CO_2 for 24 wk. The higher concentration of CO_2 stimulated plant growth, but the proportion of root length with mycorrhizae was unaffected by CO_2; results suggested that the intensity of infection was limited by the availability of root tissue for infection rather than availability of carbon supply (70). Increased CO_2 (700 ppm) stimulated the percentage of white oak (*Quercus alba* L.) root tips that were ectomycorrhizal, however (69). One conclusion from these experiments is that in an atmosphere containing twice the current ambient level of CO_2, mycorrhizae will increase either proportionally to or greater than plant growth and probably will not constrain growth of forest trees (71). This conclusion was supported by an experiment with a herbaceous plant species. In an 8-wk greenhouse experiment, subterranean clover (*Trifolium subterraneum* L.) plants were exposed to ambient (approximately 350 ppm) or 700 ppm CO_2 (24 h/day, 7 days/wk). When plants were infected by the VAM fungus *Gigaspora margarita* Becker & Hall, the higher CO_2 concentration increased final root biomass by 20% but had no effect on shoot biomass, proportion of the root system infected by the fungus, or fungal sporulation. The higher concentration of CO_2 did not affect root growth of nonVAM plants or plants infected by another VAM fungus, *Glomus intraradices* Schenck & Smith, which infected roots very weakly (3).

Exposure of plant foliage to increased concentrations of CO_2 can change the characteristics of the rhizosphere. Yellow-poplar seedlings planted in containers of a forest soil were exposed to 367 or 692 ppm CO_2. After 24 wk at the higher CO_2 concentration, the population densities (number per gram dry weight of rhizosphere soil) of nitrite-oxidizing and phosphate-dissolving bacteria were approximately half those in the rhizosphere of plants exposed to ambient CO_2 (68). Growth stimulations with increased CO_2 increase nutrient demand from soil (18), and plant biomass partitioning in elevated CO_2 may depend upon soil nutrient levels (25). As plants allocate increasing amounts or different types of carbon compounds to roots in response to an inadequate nutrient supply (14),changes in the mycorrhizal status (mycorrhizal incidence, intensity of infection, specific symbionts present) might occur.

CONSEQUENCES OF IMPACTS OF AIR POLLUTANTS ON MYCORRHIZAE

There are many examples in the literature that contain observations and data that relate mycorrhizae with the health of plants in pollution-stressed systems. These references are most common in the

literature on "forest declines" in Europe and the
eastern United States, and many have been summarized
recently (21,60). In short, many studies provide a
considerable amount of information on mycorrhizae in
stressed and unstable ecosystems. When mature trees
with reduced root biomass are subjected to
environmental stresses (drought, insect attack),
"decline" of the forest may develop (89).
Unfortunately, no study to date has established
whether the observed changes in mycorrhizal status of
trees in declining forests are among the causes or the
consequences of changing plant health and community
structure.

In many respects, the effects of various air
pollutants on mycorrhizae are not surprising. Factors
that reduce carbon allocation to roots reduce root
growth (46), and the amount of carbohydrates available
in roots in part determines mycorrhizal infection
(54). Since O_3 suppresses photosynthesis and carbon
allocation to roots, major aspects of the impact of O_3
on mycorrhizae should be similar to that induced by
shade (103), low temperature (72), or high soil
fertility (54); reduced carbon availability should
result in suppressed formation of mycorrhizae. Thus,
the suppression of mycorrhizal infection (both ecto-
and endomycorrhizae) that has been reported after
exposure of plants to O_3 seems logical.

Mycorrhizae play a role in nutrient and water
uptake, nutrient cycling, disease protection,
biological diversity, community composition, and
succession (1). Although quantification of infection
is perhaps the easiest mycorrhiza-related parameter to
measure, these measurements do not convey any
information on the extent to which a stress has
impaired the benefits of mycorrhizal infection to the
plant. Many of these benefits, aside from actual
uptake of mineral nutrients from soil by the
endophytes, are extremely important to plant fitness.
For example, organic acids released by mycorrhizae
alter nutrient availability (40). Another important
ecological effect of mycorrhizae that may be impaired
by pollutants is the mycorrhiza-induced protection
from root pathogens; increased infectious disease

following suppression of mycorrhizae is a little-studied but potentially far-reaching impact of pollutants. Defoliation of sugar maple (*Acer saccharum* Marsh.) may predispose the plants to root rot by *Armillaria mellea* Vahl ex Fr. (102), and O_3 may similarly predispose trees to root pathogens. In the field, roots of ponderosa pine and Jeffrey pine (*Pinus jiffreyi* Grew & Balf.) that exhibited severe foliar symptoms of oxidant injury were infected and colonized by *Fomes annosus* (Fr.) Cke. faster than roots of less-injured trees; in a controlled-environment experiment, the rate of pathogen growth in roots was directly related to the dose of O_3 received (39). A stress such as O_3 that alters carbon concentrations in roots probably will alter the physiological condition, and perhaps the nutrient uptake activities, of existing mycorrhizae before changes in the intensity of infection or frequencies of specific morphotypes can be measured. Thus, changes in nutrient concentrations, susceptibility of pollution-stressed plants to pathogens, and other problems with plant health attributed to pollutants directly theoretically could be mediated through changes in mycorrhizal function.

The same "common biological sense (that) indicates that root symbiotic associations are responsive to effects of atmospheric deposition" (55) does not, unfortunately, tell us what to expect in terms of the quantitative response of plant communities and ecosystems to mycorrhiza-mediated pollution stress. The impact on plant and ecosystem fitness remains unknown because the experimental constraints under which most experiments are conducted simply do not permit an assessment of this type (87). That air pollution can play a role with the consequences of mycorrhizal infection can be suggested, however, by consideration of seemingly unrelated experiments. In an experiment on effects of chronic O_3 on competition between crimson clover and annual ryegrass, a rather common level of tropospheric O_3 (90 ppb) changed the community balance in the plots in favor of the ryegrass (*Secale cereale* L.) (5); this finding was supported by another experiment with ladino clover (*Trifolium rapens* L.) and tall fescue (*Festuca elatior* L.) (32,74). In general, clovers are more mycorrhizal-dependent than grasses (13), due perhaps to differences in root architecture or the presence of the additional root symbiosis (i.e., rhizobium nodules) on clover. Thus, mycorrhizae may be playing a role in the O_3-induced shifts in these simple mixtures of plant species, but this remains conjecture in the absence of studies designed to study this specific possibility. Implications for complex natural ecosystems can only be suggested. One possible consequence of chronic pollution stress, for example, is a shift from a community dominated by ectomycorrhizal plant species to one dominated by endomycorrhizal species as trees decline and herbaceous plants, including grasses, continue to flourish (101).

WHERE DO WE GO FROM HERE?

A philosophical dilemma that has been expressed for scientific research in general is certainly germane to mycorrhiza researchers, especially to mycorrhiza researchers working in the air pollution field in particular. Sir Peter Medawar characterized science as "the art of the soluble", and this concept was elaborated by Loehle (51) to mean that working on easily-solved problems represents little if any scientific advance, whereas attacking extremely difficult scientific problems may be exceptionally risky because current technology or scientific understanding may prevent success. Thus, "intermediate problems have the highest benefit per unit of effort because they are neither too simple to be useful nor too difficult to be solvable" (51). We face these basic concepts in the conduct of research on the role of mycorrhizae in mediating plant, community, and ecosystem response to pollution. It is relatively easy to expose plants to a pollutant or other stress and quantify infection at the end of the experiment; for awhile, this approach represented a scientific advance because effects of pollutants on mycorrhizae had been essentially unknown. Such information told us that here was a phenomenon worth investigating; it now offers little significant new insight. On the other hand, technical and sampling constraints occur in air pollution studies. These include controlled environments (pots, chambers, artificial pollution exposure dynamics) and their unavoidable artifacts that limit extrapolation to the outside world, and limitations in sampling (numbers and/or repeated in time) because of a restricted number of plants contained within the chambers. These constraints, associated with an imperfect quantitative understanding of the impact of altered mycorrhizal infection on plants and plant communities, render many really important questions currently beyond "the art of the soluble." The most creative work in this field will come from those individuals who respond to quantification of mycorrhizae following exposure of plants to pollutant stress with research regarding the implications of observed changes in infection, somehow keeping the effort within Loehle's (51) "Medawar Zone." Unfortunately, this work may not be very attractive to granting agencies because the specific outcome is not predictable. We can be sure, however, that much of this research will fall into the area of host-endophyte physiology; O'Neill et al. (71) correctly observed that, in the context of air pollution research, "mycorrhizal researchers are faced with the dilemma of designing experiments on small-scale processes that contribute to the solution of large-scale problems", and that it is "physiological processes that are the fundamental mechanisms underlying large-scale ecosystem behavior."

Even though there are limitations in our current knowledge and technical and philosophical dilemmas to be faced, some alternative approaches to investigating interactions of air pollutants, plants, and mycorrhizal fungi are worth considering. Considerable effort should be shifted from the emphasis on "responses of mycorrhizae to pollution" to an emphasis on "mechanisms by which mycorrhizae mediate plant response to pollution." Following answers to some critical questions in that regard, an emphasis on "consequences of pollution-mycorrhiza interactions for communities and ecosystems" can be evaluated. Indeed, the supposition here is that the pollution-mycorrhiza interactions are of sufficient magnitude to be important on an community or ecosystem level; evaluation of this supposition itself has been debated in the context of the needs of assessing a global impact of climate change (71).

A major question that needs to be addressed is whether the observed magnitudes of responses of mycorrhizae to air pollution have any consequences for the fitness of an individual plant. Rhetorical questions heard (and raised) by almost anyone who has worked with mycorrhizae to any extent include "What is the functional significance of intense infection versus sparse infection?" "Are numerically equivalent infection intensities by different fungi functionally equivalent?" "How do many short extraradicle hyphae relate to a few long hyphae in terms of function and plant response?" These are examples of basic questions in the field of mycorrhiza studies that remain unanswered in large part, yet they are critical to evaluation of the impact of pollutants. Approach to such topics in the context of air pollution studies would serve many interests at once.

Questions related to the fitness of plants in a polluted environment should include a mycorrhizal component. Leguminous plants, for example, represent a three-part symbiosis of a carbon-fixing plant, the nitrogen-fixing rhizobia, and the P nutrient-gathering VAM fungi. An experiment with subterranean clover illustrates the importance of evaluating symbiont function to assess the impact of a pollutant on a plant (88). Over an 8-wk period, subterranean clover plants were exposed to a graded series of O_3 concentrations in greenhouse chambers. Root growth but not shoot growth was suppressed by 100 and 150 ppb O_3 relative to that of plants exposed to 0 or 50 ppb O_3. Symbiotic microorganisms in the roots apparently were affected by the suppressed carbon availability. This was evident not as suppressed nodulation or mycorrhizal infection, which were unaffected; rather, the function of the nodules in terms of nitrogen fixation was lessened during the course of the study.

Analyses of ^{15}N, which was absorbed from labeled
sources in soil, in the plant tissues indicated that
O_3-stressed plants relied on soil nitrogen pools more
than nonstressed plants did. Thus, the implications
for plant fitness or persistence in the field include
not only the impact on root growth, but lowered
nitrogen availability in the soil in subsequent years
(88). Understanding the impact of pollutants on plant
communities will be advanced much further by work that
examines the role of root symbionts in aspects of
plant function than by more studies that quantify
effects of pollutants on numbers of nodules or
mycorrhizal infection.

The inconsistencies that have occurred among
studies of pollution effects and mycorrhizae simply
reflect the complexity of the system in which the
symbiotic association is but a part. Although a
difficult task, understanding the importance of
mycorrhizae in ecosystem response to pollutants will
require "dissection" of the important environmental
and biological interactions. Of course, the
importance of assorted environmental influences on
mycorrhizae are known (12). To date, the major
dependent variable measured for mycorrhizae in
pollution studies (i.e., infection quantification)
represents the endpoint result of many interacting
factors. One example of the complexity of these
interactions is the genetic variation in plant
responses to a pollutant. Within a species,
considerable variation exists with respect to response
to O_3; this has been demonstrated for a wide variety
of plants, including both endomycorrhizal and
ectomycorrhizal species (7,30,85; studies reviewed in
33). Moreover, the identity of the endophyte also
must be considered; O_3 effects on tomato plants
infected by different species of VAM fungi were not
uniform (57). Soil or growth medium is another
variable that conditions plant growth, mycorrhizal
formation and/or function, plant response to a
pollutant, and mycorrhizal response to a pollutant
(7,35). Drought stress alters plant response to O_3
(29,31,32). Maintenance respiration by roots
increases with drought stress (11); thus, plants

stressed by both O_3 and drought hypothetically could exhibit rapid depletion of soluble carbohydrates in roots, followed by a decrease in mycorrhizal infection, formation, or function. A study of this possibility, however, indicated no interaction between O_3 and drought, and only the main-effect of O_3 suppressed mycorrhizal frequency (61).

Models represent one approach to understanding the integrated, complex systems of plants, mycorrhizal fungi, other symbionts, other microorganisms, soil characteristics, and many environmental variables. Mathematical models provide the capability to conceptually manipulate systems and test hypothetical influences on a system. Models may offer several advantages, such as "testing" an effect over temporal and spatial scales currently untestable by other empirical means. The types of models, their inherent advantages and disadvantages, and their utility in air pollution work have been considered elsewhere (44). Other models are being developed for assessing atmospheric pollution impact at the various scales of concern (23). Unfortunately, mycorrhizae are rarely a discrete component in these models. In one conceptual model (6), mycorrhiza was a discrete component integrated into the level of the whole-plant study. Although conceptually incorporated into this multidisciplinary effort, the mycorrhizal data (as well as other components) were not easily integrated to provide a unified explanation of pollutant impact on plant function. The authors (7) lamented the fact that even though the number of participating groups was large, most studies simply complemented each other without overlap; furthermore, most studies were restricted to analyses at the end, so interpretation in the context of nondestructive plant analyses conducted throughout the experiment was difficult. These inequities in approach led to disparate data sets. The data on mycorrhizae, in particular, provided only a point-in-time quantification of infection itself that had limited value in providing an understanding of the feedbacks among the components in the conceptual model.

Mycorrhiza researchers should be concerned that

mycorrhizae are rarely specific components of models for pollution-stressed plant systems and should wonder about the value our data has to others. Why do other plant scientists and ecologists often ignore our findings? Do we poorly promote our useful information to those outside our specific field? Is the mycorrhizal component important enough in driving a system that it needs to be included in pollution-stress models anyway? If it is, are the data we gather appropriate for use in the models, i.e., does the information provide a predictive capability? Or are we gathering data to suit only our own short-term goals or curiosity? One problem of utilizing the mycorrhizal data collected to date in modelling efforts that seek to address the larger-scales of concern (i.e. whole-plant, ecosystem, landscape, etc.) is related to the differences in scales (1,71). Whether or not information on mycorrhizae can add value to the predictive capability of a model is being addressed by the U. S. Environmental Protection Agency.   Parameters in the model that can be changed directly by mycorrhizae (i.e., functional radius of the root, kinetics of nutrient uptake) are identified, values of these parameters are obtained from existing data on mycorrhizae and entered, and the impact of the mycorrhizae on the outcome of the model is assessed (C. P. Andersen, US-EPA, Corvallis, OR, personal communication). This approach may provide insight into whether mycorrhizae should be included as a discrete component in these models, whether modifications of parameters for existing components (i.e., coarse roots, fine roots) to incorporate the mycorrhizal attributes will suffice, or whether mycorrhizal information is needed at all in large-scale questions related to climate change.

CONCLUSIONS

Effects of major air pollutants on plants have been studied intensively for decades. The information that is available leads to the logical conclusion that the impacts of pollutants on carbon partitioning, rhizosphere characteristics, and nutrient balances within plants and between plants and soils will alter relationships between plants and mycorrhizal fungi. Research over the past 20 years has demonstrated that this is the case for both ecto- and endomycorrhizal systems. The implications for these impacts on plant ecology remain conjecture because efforts to date have not ventured far beyond the quantification of infection itself. Models that are designed to predict the impact of pollutants on plant communities and ecosystems may provide some insight into the importance of mycorrhizae in ecosystem responses to pollutants, but these models are only as good as the data that drive them. Quantitative descriptions of relationships among environmental stresses, plant physiology, mycorrhizal infection, and mycorrhizal function will be needed before the importance of mycorrhizae in pollution-related changes in the ecosystem can be assessed.

LITERATURE CITED

1.   Allen, M.F. 1992. The Ecology of Mycorrhizae. Cambridge University Press, New York, NY.
2.   Arnolds, E. 1991. Decline of ectomycorrhizal fungi in Europe. Agric., Ecosystems Environ. 35: 209-244.
3.   Bamford, M. S. 1992. Impact of ozone and carbon dioxide on plant biomass and mycorrhizal development of subterranean clover. M.S. Thesis, North Carolina State University. 61 pp.
4.   Bazzaz, F. A. 1990. The response of natural ecosystems to the rising global CO_2 levels. Ann. Rev. Ecol. Syst. 21: 167-196.
5.   Bennett, J. P., and Runeckles, V. C. 1977. Effects of low levels of ozone on plant competition. J. Appl. Ecol. 14: 877-880.
6.   Blank, L.W., Payer, H.D., Pfirrmann, T., Gnatz, G., Kloos, M., Runkel, K.-H., Schmolke, W., Strube, D., and Rehfuess, K.E. 1990. Effects of ozone, acid mist and soil characteristics on clonal Norway spruce (*Picea abies* (L.) Karst.) - An introduction to the joint 14 month tree

exposure experiment in closed chambers. Env.
Poll. 64:189-207.

7. Blank, L.W., Payer, H.D., Pfirrmann, T., and
Rehfuess, K.E. 1990. Effects of ozone, acid mist
and soil characteristics on clonal Norway spruce
(*Picea abies* (L.) Karst.) - Overall results and
conclusions of the joint 14 month tree exposure
experiment in closed chambers. Env. Poll.
64:385-395.

8. Blum, U., Mrozek, E., Jr., and Johnson, E.
1983. Investigation of ozone (O_3) effects on
14-C distribution in ladino clover. Environ.
Exp. Bot. 23: 369-378.

9. Blum, U., and Tingey, D. T. 1977. A study of
the potential ways in which ozone could reduce
root growth and nodulation of soybeans. Atmos.
Environ. 11: 737-739.

10. Brewer, P.F., and Heagle, A.S. 1983.
Interactions between *Glomus geosporum* and
exposure of soybean to ozone or simulated acid
rain in the field. Phytopathology 73:1035-1040.

11. Brix, H. 1962. The effects of water stress on
the roles of photosynthesis and respiration in
tomato plants and loblolly pine seedlings.
Physiol. Plant. 15:10-20.

12. Brundrett, M. 1991. Mycorrhizas in natural
ecosystems. Advances in Ecological Research 21:
171-313.

13. Buwalda, J. G. 1980. Growth of a
clover-ryegrass association with vesicular
arbuscular mycorrhizas. N. Z. J. Agric. Res.
23: 379-383.

14. Chapin, F. S., III. 1991. Integrated responses
of plants to stress. BioScience 41: 29-36.

15. Chevone, B.I., and Linzon, S.N. 1988. Tree
decline in North America. Environ. Pollut. 50:
87-99.

16. Cooley, D. R., and Manning, W. J. 1987. The
impact of ozone on assimilate partitioning in
plants: a review. Environ. Pollut. 47: 95-113.

17. Coyne, P. I., and Bingham, G. E. 1981.
Comparative ozone dose response of gas exchange
in a ponderosa pine stand exposed to long-term

fumigations. J. Air Pollut. Cont. Assoc. 31: 38-41.

18.   Cure, J. D. 1985. Carbon dioxide doubling responses: a crop survey. Pages 99-116 in: Direct Effects of Increasing Carbon Dioxide on Vegetation. B. R. Strain and J. D. Cure, eds., DOE/ER-0238. U. S. Department of Energy, Washington, D.C.

19.   Cwyndar, T. 1992. Fewer fungi bode poorly for forests. Missouri Conservationist 53: 28.

20.   Del Castillo, D., Acock, B., Reddy, V. R., and Acock, C. 1989. Elongation and branching of roots on soybean plants in a carbon dioxide-enriched aerial environment. Agron. J. 81: 692-695.

21.   Dighton, J., and Jansen, A. E. 1991. Atmospheric pollutants and ectomycorrhizae: more questions than answers? Environ. Pollut. 73: 179-204.

22.   Dighton, J., and Skeffington, R. A. 1987. Effects of artificial acid precipitation on the mycorrhizas of Scots pine seedlings. New Phytol. 107: 191-202.

23.   Dixon R.K., Meldahl, R.S., Ruark, G.A. and Warren, W.G. (eds.) 1990. Process modeling of forest growth responses to environmental stress. Timber Press, Portland, OR.

24.   Duponnois, R., and Garbaye, J. 1991. Effect of dual inoculation of Douglas fir with the ectomycorrhizal fungus *Laccaria laccata* and mycorrhization helper bacteria (MHB) in two bare-root forest nurseries. Plant Soil 138: 169-176.

25.   El Kohen, A., Rouhier, H., and Mousseau, M. 1992. Changes in dry weight and nitrogen partitioning induced by elevated CO_2 depend on soil nutrient availability in sweet chestnut (*Castanea sativa* Mill). Ann. Sci. For. 49: 83-90.

26.   Evans, L. S., Curry, T. M., and Lewin, K. F. 1981. Responses of leaves of *Phaseolus vulgaris* to simulated acid rain. New Phytol. 88: 403-420.

27.   Evans. L. S., Gmur, N. F., and DaCosta, F.
       1977. Leaf surface and histological
       perturbations of leaves of *Phaseolus vulgaris*
       and *Helianthus annuus* after exposure to
       simulated acid rain. Amer. J. Bot. 64: 903-913.
28.   Gorissen, A., Joosten, N.N., and Jansen, A.E.
       1991. Effects of ozone and ammonium sulfate on
       carbon partitioning to mycorrhizal roots of
       juvenile Douglas-fir. New Phytol. 119:243-250.
29.   Heagle, A. S., Flagler, R. B., Patterson, R. P.,
       Lesser, V. M., Shafer, S. R., and Heck, W. W.
       1987. Injury and yield response of soybean to
       chronic doses of ozone and soil moisture
       deficit. Crop Sci. 27: 1016-1024.
30.   Heagle, A. S., McLaughlin, M. R., Miller, J. E.,
       Joyner, R. L., and Spruill, S. E. 1991.
       Adaptation of a white clover population to ozone
       stress. New Phytol. 119: 61-68.
31.   Heagle, A. S. Miller, J. E., Heck, W. W., and
       Patterson, R. P. 1988. Injury and yield
       response of cotton to chronic doses of ozone and
       soil moisture deficit. J. Environ. Qual. 17:
       627-635.
32.   Heagle, A. S., Rebbeck, J., Shafer, S. R.,
       Lesser, V. M., Blum, U., and Heck, W. W. 1989.
       Effects of long-term O_3 exposure and soil
       moisture deficit on growth of a white
       clover-tall fescue pasture. Phytopathology 79:
       128-136.
33.   Heck, W. W., Heagle, A. S., and Shriner, D. S.
       1986. Effects on vegetation: native, crops,
       forests. Pages 247-350 in: Air Pollution,
       Volume 6. A. S. Stern, ed., Academic Press,
       New York.
34.   Hetrick, B. A. D., Wilson, G. W. T., and Todd,
       T. C. 1990. Differential responses of C3 and
       C4 grasses to mycorrhizal symbiosis, phosphorus
       fertilization, and soil microorganisms. Can. J.
       Bot. 68: 461-467.
35.   Horton, S. J., Reinert, R. A., and Heck, W. W.
       1990. Effects of ozone on three open-pollinated
       families of *Pinus taeda* L. grown in two
       substrates. Environ. Pollut. 65: 279-292.

36. Idso, S. B., Kimball, B. A., and Mauney, J. R. 1988. Effects of atmospheric CO_2 enrichment on root:shoot ratios of carrot, radish, cotton, and soybean. Agric., Ecosystems Environ. 21: 293-299.

37. Irving, P. M. 1983. Acidic precipitation effects on crops: a review and analysis of research. J. Environ. Qual. 12: 442-453.

38. Irving, P. M. (ed.). 1991. Acidic Deposition: State of Science and Technology, Volumes I-IV. National Acidic Precipitation Assessment Program, Washington, DC.

39. James, R. L., Cobb, F. W., Jr., Miller, P. R., and Parmeter, J. R., Jr. 1980. Effects of oxidant air pollution on susceptibility of pine roots to *Fomes annosus*. Phytopathology 70: 560-563.

40. Jayachandran, K. Schwab, A.P., and Hetrick, B.A.D. 1989. Mycorrhizal mediation of phosphorus availability: synthetic iron chelate effects on phosphorus solubilization. Soil Sci. Soc. Am. J. 53:1701-1706.

41. Johnston, J. W., Jr., Shriner, D. S., Klarer, C. I., and Lodge, D. M. 1982. Effect of rain pH on senescence, growth, and yield of bush bean. Environ. Exp. Bot. 22: 329-337.

42. Keane, K. D., and Manning, W. M. 1988. Effects of ozone and simulated acid rain on birch seedling growth and formation of ectomycorrhizae. Environ. Pollut. 52: 55-65.

43. Kickert, R. N., and Krupa, S. V. 1989. The greenhouse effect: impacts of ultraviolet-B (UV-B) radiation, carbon dioxide (CO_2), and ozone (O_3) on vegetation. Environ. Pollut. 61: 263-393.

44. Kickert, R. N., and Krupa, S. V. 1991. Modeling plant response to tropospheric ozone: a critical review. Environ. Pollut. 70: 271-383.

45. Koziol, M. J., and Whatley, F. R. (eds.). 1984. Gaseous Air Pollutants and Plant Metabolism. Butterworths, London. 466 pp.

46. Larsen, H. S., South, D. B., and Williams, H. M. 1989. Pine seedling root growth is reduced by

defoliation and shading. Alabama Agric. Exp.
Sta. 36: 14.

47. Lechowicz, M. J. 1987. Resource allocation by
plants under air pollution stress: implications
for plant-pest-pathogen interactions. Bot. Rev.
53: 281-300.

48. Lefohn, A. S. (ed.). 1992. Surface Level Ozone
Exposures and Their Effects on Vegetation.
Lewis Publishers, Chelsea, MI 366 pp.

49. Lefohn, A. S., Krupa, S. V., and Winstanley, D.
1990. Surface ozone exposures measured at clean
locations around the world. Environ. Pollut. 63:
189-224.

50. Leone, I. A., and Brennan, E. 1970. Ozone
toxicity in tomato as modified by phosphorus
nutrition. Phytopathology 60: 1521-1524.

51. Loehle, C. 1990. A guide to increased
creativity in research - Inspiration or
perspiration? BioScience 40: 123-129.

52. Mahoney, M. J., Chevone, B. I., Skelly, J. M.,
and Moore, L. D. 1985. Influence of
mycorrhizae on the growth of loblolly pine
seedlings exposed to ozone and sulfur dioxide.
Phytopathology 75: 679-682.

53. Manning, W. J., Feder, W. A., Papia, P. M., and
Perkins, I. 1971. Influence of foliar ozone
injury on root development and root surface
fungi of pinto bean plants. Environ. Pollut.
11: 305-312.

54. Marx, D. H., Hatch, A. B., and Mendicino, J. F.
1977. High soil fertility decreases sucrose
content and susceptibility of loblolly pine
roots to ectomycorrhizal infection by *Pisolithus
tinctorius*. Can. J. Bot. 55: 1569-1574.

55. Marx, D. H., and Shafer, S. R. 1989. Fungal
and bacterial symbioses as potential biological
markers of effects of atmospheric deposition on
forest health. Pages 217-232 in: Biologic
Markers of Air Pollution Stress and Damage in
Forests. National Academy Press, Washington, D.
C.

56. McCool, P.M., and Menge, J.A. 1983. Influence of
ozone on carbon partitioning in tomato:

potential role of carbon flow in regulation of the mycorrhizal symbiosis under conditions of stress. New Phytol. 94: 241-247.

57. McCool, P.M., and Menge, J.A. 1984. Interaction of ozone and mycorrhizal fungi on tomato influenced by fungal species and host variety. Soil Biol. Biochem 16: 425-427.

58. McCool, P.M., Menge, J.A., and Taylor, O.C. 1982. Effect of ozone injury and light stress on response of tomato to infection by the vesicular-arbuscular mycorrhizal fungus, *Glomus fasciculatus*. J. Amer. Soc. Hort. Sci. 107: 839-842.

59. McLaughlin, S. B. 1985. Effects of air pollution on forests, a critical review. J. Air Pollut. Cont. Assoc. 35: 512-534.

60. Meier, S. 1991. Quality versus quantity: optimizing evaluation of ectomycorrhizae for plants under stress. Environ. Pollut. 73: 205-216.

61. Meier, S., Grand, L. F., Schoeneberger, M. M., Reinert, R. A., and Bruck, R. I. 1990. Growth, ectomycorrhizae and nonstructural carbohydrates of loblolly pine seedlings exposed to ozone and soil water deficit. Environ. Pollut. 64: 11-27.

62. Meier, S., Robarge, W. P., Bruck, R. I., and Grand, L. F. 1989. Effects of simulated rain acidity on ectomycorrhizae of red spruce seedlings potted in natural soil. Environ. Pollut. 59: 315-324.

63. Meyer, J. R., and Linderman, R. G. 1986. Response of subterranean clover to dual inoculation with vesicular-arbuscular mycorrhizal fungi and a plant growth-promoting bacterium, *Pseudomonas putida*. Soil Biol. Biochem. 18: 185-190.

64. Miller, J. E. 1987. Effects of ozone and sulfur dioxide stress on growth and carbon allocation in plants. Pages 55-100 in: Recent Advances in Phytochemistry: Volume 21, Phytochemical Effects of Environmental Compounds. J. A. Saunders, L. Kosak-Channing, and E. E. Conn, eds., Plenum Press, New York.

65.   Miller, P. R., Longbotham, G. J., and
       Longbotham, C. R. 1983. Sensitivity of
       selected western conifers to O_3. Plant Dis. 67:
       1113-1115.
66.   Mooney, H. A., and Winner, W. E. 1988. Carbon
       gain, allocation, and growth as affected by
       atmospheric pollutants. Pages 272-287 in: Air
       Pollution and Plant Metabolism. S.
       Schulte-Hostde, N. M. Darrall, L. W. Blank, and
       A. R. Wellburn, eds., Elsevier Applied Science,
       London.
67.   Morton, J. B. 1990. Evolutionary relationships
       among arbuscular mycorrhizal fungi in the
       Endogonaceae. Mycologia 82: 192-207.
68.   O'Neill, E. G., Luxmoore, R. J., and Norby, R.
       J. 1987. Elevated atmospheric CO_2 effects on
       seedling growth, nutrient uptake, and
       rhizosphere bacterial populations of
       Liriodendron tulipifera L. Plant Soil 104:
       3-11.
69.   O'Neill, E. G., Luxmoore, R. J., and Norby, R.
       J. 1987. Increases in mycorrhizal colonization
       and seedling growth in *Pinus echinata* and
       Quercus alba in an enriched CO_2 atmosphere.
       Can. J. For. Res. 17: 878-883.
70.   O'Neill, E. G., and Norby, R. J. 1988.
       Differential responses of ecto- and
       endomycorrhizae to elevated atmospheric CO_2.
       Bull. Ecol. Soc. Amer. Supp. 69: 248-249.
71.   O'Neill, E. G., O'Neill, R. V., and Norby, R. J.
       1991. Heirarchy theory as a guide to
       mycorrhizal research on large-scale problems.
       Environ. Pollut. 73: 271-284.
72.   Parke, J. L., Linderman, R. G., and Trappe, J.
       M. 1983. Effect of root zone temperature on
       ectomycorrhiza and vesicular-arbuscular
       mycorrhiza formation in disturbed and
       undisturbed forest soils of southwest Oregon.
       Can J. For. Res. 13: 657-665.
73.   Radin, J. W., Parker, L. L., and Sell, C. R.
       1978. Partitioning of sugar between growth and
       nitrate reduction in cotton roots. Plant
       Physiol. 62: 550-553.

74. Rebbeck, J., Blum, U., and Heagle, A. S. 1988. Effects of ozone on the regrowth and energy reserves of a ladino clover-tall fescue pasture. J. Applied Ecology 25: 659-681.

75. Reich, P. B., and Amundson, R. G. 1985. Ambient levels of ozone reduce net photosynthesis in tree and crop species. Science 230: 566-570.

76. Reich, P. B., Schoettle, A. W., Stroo, H. F., and Amundson, R. G. 1986. Acid rain and ozone influence mycorrhizal infection in tree seedlings. J. Air Pollut. Cont. Assoc. 36: 724-726.

77. Reich, P. B., Schoettle, A. W., Stroo, H. F., Troiano, J., and Amundson, R. G. 1985. Effects of O_3, SO_2 and acidic rain on mycorrhizal infection in northern red oak seedlings. Can. J. Bot. 63: 2049-2055.

78. Reinert, R. A. 1984. Plant response to air pollutant mixtures. Annu. Rev. Phytopathol. 22: 421-442.

79. Schroeder, P., and Kiester, A. R. 1989. The Forest Response Program: national research on forest decline and air pollution. J. Forestry 87: 27-32.

80. Schutt, P., and Cowling, E. B. 1985. Waldsterben, a general decline of forests in central Europe: symptoms, development and possible causes. Plant Dis. 69:548-558.

81. Shafer, S. R. 1988. Influence of ozone and simulated acidic rain on microorganisms in the rhizosphere of Sorghum. Environ. Pollut. 51: 131-152.

82. Shafer, S. R. 1992. Responses of microbial populations in the rhizosphere to deposition of simulated acidic rain onto foliage and/or soil. Environ. Pollut. 76: 267-278.

83. Shafer, S. R., Bruck, R. I., and Heagle, A. S. 1985. Influence of simulated acidic rain on *Phytophthora cinnamomi* and Phytophthora root rot of blue lupine. Phytopathology 75: 996-1003.

84. Shafer, S. R., Grand., L. F., Bruck, R. I., and Heagle, A. S. 1985. Formation of

ectomycorrhizae on *Pinus taeda* seedlings exposed to simulated acidic rain. Can J. For. Res. 15: 66-71.

85.   Shafer, S. R., and Heagle, A. S. 1989. Growth responses of field-grown loblolly pine to chronic doses of ozone during multiple growing seasons. Can. J. For. Res. 19: 821-831.

86.   Shafer, S. R., Koenning, S. R., and Barker, K. R. 1992. Interactions of simulated acidic rain with root-knot or cyst nematodes on soybean. Phytopathology 82: 962-970.

87.   Shafer, S. R., and Schoeneberger, M. M. 1991. Mycorrhizal mediation of plant response to atmospheric change: air quality concepts and research considerations. Environ. Pollut. 73: 163-177.

88.   Shafer, S. R., and Schoeneberger, M. M. 1991. Ozone-induced alteration of biomass allocation and nitrogen derived from air in the leguminous plant-Rhizobium-VAM system. Page 377 in: The Rhizosphere and Plant Growth. D. L. Keister and P. B. Cregan, eds., Kluwer Academic Publishers, The Netherlands.

89.   Sharpe, P. J. H., and Scheld, H. W. 1986. The role of mechanistic modeling in estimating long-term pollution effects upon natural and man-influenced forest ecosystems. Pages 72-82 in: Proceedings of a Workshop on Controlled Exposure Techniques and Evaluation of Tree Responses to Airborne Chemicals. NCASI Technical Bulletin, No. 500, August 1986.

90.   Sheffield, R. M., Cost, N. D. Bechtold, W. A., and McClure, J.P. 1985. Pine growth reductions in the southeast. USDA-Forest Service Resources Bulletin SE-83, Southeastern Forest Experiment Station, Asheville, NC. 112 pp.

91.   Siccama, T. G., Bliss, M., and Vogelmann, H. W. 1982. Decline of red spruce in the Green Mountains of Vermont. Bull. Torrey Bot. Club 109: 163.

92.   Simmons, G. L., and Kelly, J. M. 1989. Influence of O_3, rainfall acidity, and soil Mg status on growth and ectomycorrhizal

colonization of loblolly pine roots. Water, Air, Soil Pollut. 44: 159-171.

93. Smith, J. B., and Tirpak, D. 1989. The Potential Effects of Global Climate Change on the United States: Report to Congress. EPA-230-05-89-050. U.S. Environ. Prot. Agency, Washington, DC. 413 pp + appendices.

94. Spence, R. D., Rykiel, E. J., Jr., and Sharpe, P. J. H. 1990. Ozone alters carbon allocation in loblolly pine: assessment with carbon-11 labeling. Environ. Pollut. 64: 93-106.

95. Strain, B. R., and Cure, J. D. 1985. Direct Effects of Increasing Carbon Dioxide on Vegetation. DOE/ER-0238, U.S. Dept.of Energy, Washington, DC. 286 pp.

96. Stroo, H. F., and Alexander, M. 1985. Effect of simulated acid rain on mycorrhizal infection of *Pinus strobus* L. Water, Air, Soil Pollut. 25: 107-114.

97. Stroo, H. F., Reich, P. B., Schoettle, A. W., and Amundson, R. G. 1988. Effects of ozone and acid rain on white pine (*Pinus strobus*) seedlings grown in five soils. II. Mycorrhizal infection. Can. J. Bot. 66: 1510-1516.

98. Turner, N. C., Rich, S., and Waggoner, P. E. 1973. Removal of ozone by soil. J. Environ. Qual. 2: 259-264.

99. Unsworth, M. H., and Ormrod, D. P. (eds.). 1982. Effects of Gaseous Air Pollution in Agriculture and Horticulture. Butterworth Scientific, London. 532 pp.

100. Visser, S., Danielson, R. M., and Parr, J. F. 1987. Effects of Acid-Forming Emissions on Soil Microorganisms and Microbially-Mediated Processes. Prepared for the Acid Deposition Research Program by the Kannaskis Centre for Environmental Research, The University of Calgary, and U. S. Department of Agriculture, Beltsville, Maryland. ADRP-B-02-87. 86 pp.

101. Vosatka, M., Cudlin, P., and Mejstrik, V. 1991. VAM populations in relation to grass invasion associated with forest decline. Environ. Pollut. 73: 263-270.

102.   Wargo, P. M.   1972.   Defoliation-induced
        chemical changes in sugar maple roots stimulate
        growth of *Armillaria mellea*.   Phytopathology 62:
        1278-1283.
103.   Wright, E. 1971. Mycorrhizae on Douglas-fir and
        Ponderosa pine seedlings. Research Bulletin #13.
        Forest Research Laboratory, Oregon St. Univ.,
        Corvallis. 36 pp.
104.   World Resources Institute.   1990.   World
        Resources 1990-91.   Oxford University Press, New
        York.   383 pp.

VESICULAR-ARBUSCULAR MYCORRHIZAE AND BIOGEOCHEMICAL CYCLING

R. M. Miller and J. D. Jastrow
Argonne National Laboratory
Argonne, Illinois

An ecosystem may be defined as a biogeochemical system in which persistent organic structures are formed within a surrounding inert geochemical matrix (44). In this context, the fluxes and cycling of carbon and nutrients between an ecosystem's biotic components and its geochemical matrix are fundamental properties that define or control ecosystem behavior. At the ecosystem level of organization, the importance of the mycorrhizal association is derived from the fact that mycorrhizae form a fundamental link between the biotic and geochemical portions of the system (43). Furthermore, mycorrhizae perform this coupling role for practically all terrestrial ecosystems.

The state, concentration, and distribution of nutrient ion pools vary among soils depending upon the soil-forming factors of vegetation, parent material, topography, climate, and time (26). These same factors not only determine the contributions of mycorrhizae to biogeochemical cycling, but they also determine the types of mycorrhizae that predominate in a system. The large scale processes that delineate biomes also appear to determine whether the system is composed of ectomycorrhizal fungi, ericoid or arbutoid mycorrhizal fungi, or vesicular-arbuscular mycorrhizal (VAM) fungi. Read (46) has eloquently argued that a specific suite of climatic and edaphic conditions have led to the evolution of distinctive types of mycorrhizae that are biome related. For example, VAM predominate in herbaceous and woody plant ecosystems on mineral soils, while ericoid mycorrhizae predominate on more humus soils, and ectomycorrhizae predominate in forest ecosystems with surface litter accumulation. Read's global-scale analyses suggest that climate, through its effect on pedogenic processes, is the primary determinant influencing the type of mycorrhizal community (Figure 1). At much finer scales, however, he emphasizes that the quality of soil nutrient resources is of primary importance in determining the kinds of mycorrhizae present, and that climate is at most a secondary factor. Thus at a local level, soil humus type, the C:N ratio of litter, and soil pH are major determinants of mycorrhizal type (46).

In this chapter, we focus our discussion on the contributions of VAM to biogeochemical cycling. In considering the VAM association, the complexity of

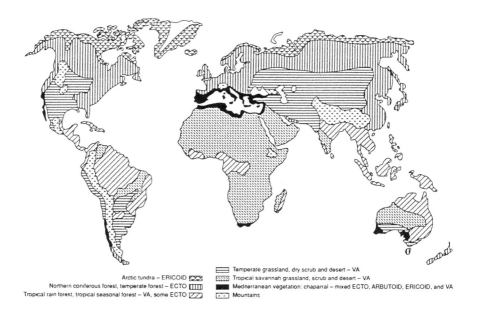

Arctic tundra – ERICOID
Northern coniferous forest, temperate forest – ECTO
Tropical rain forest, tropical seasonal forest – VA, some ECTO

Temperate grassland, dry scrub and desert – VA
Tropical savannah grassland, scrub and desert – VA
Mediterranean vegetation: chaparral – mixed ECTO, ARBUTOID, ERICOID, and VA
Mountains

Figure 1. The distribution of major terrestrial biomes of the world, demonstrating the large-scale relationship between identifiable systems and mycorrhizal type (redrawn from Read [46]).

responses expressed by both plants and fungi to their environment must be recognized, because many of these responses involve fundamental mechanisms underlying ecosystem behavior (Table 1). Both plant and VAM fungal responses are exhibited in their physiology and life history traits. However, rather than host and fungus acting in isolation, their responses are mediated by each other's activities. For example, fungal traits such as the kinds of propagules produced and the allocation of nutrients for growth between intraradical and extraradical hyphae are influenced by local factors but can also be affected by host responses such as adjustments in photosynthesis, biomass allocation, and root morphology. In addition, the mycorrhizal association, i.e. roots and fungus, may also respond to adjustments by neighboring species, which further emphasizes the complexity of interactions that may result in host or fungus responses.

The contributions of the mycorrhizal association at larger organizational scales are indirect. At the community level, the mycorrhizal association appears to influence community structure via mutualistic and competitive processes. However, at the ecosystem level, process level changes affected by the mycorrhizal association are an indirect result of mycorrhizal influences on vegetation patterns within the community (e.g., succession).

Although studies often involve variations in available soil nutrients, a neglected area of mycorrhizal research is the integration of plant and VAM fungal responses with biogeochemical and nutrient cycling. One reason for this neglect is that plant and fungal responses and biogeochemical cycling are often studied at different scales (3,43). From a conceptual viewpoint, nevertheless, VAM fungal hyphae do contribute to system processes and functions at various hierarchical organizational levels (Figure 2). Thus, linkages and feedbacks between mycorrhizae and biogeochemical cycling exist. The difficulty lies in our ability to focus questions and to measure responses or processes that function as control points for biogeochemical cycling.

Table 1. Examples of direct and indirect expression of the mycorrhizal association on processes occurring at various hierarchical organization levels of system structure.

Structural organization	Processes and phenomenon
Mycorrhizal roots,	Ion uptake, organic acid production, external hyphae, soil aggregation, root morphology
	Whole plant nutrition, growth, water relations, disease susceptibility
Community	Diversity, stability, competition
Ecosystem	Nutrient retention, conservative cycling, positive feedback

Our current knowledge of the contribution of VAM fungi to biogeochemical cycling is limited both conceptually and experimentally. Although many of the interactions between plants and mycorrhizal fungi in relation to nutrient cycling are at lower levels of organization, the outcome of these interactions potentially can influence large-scale, ecosystem level responses (3,43,45). Most mycorrhizasts, however, have avoided the difficulties of conducting large-scale field experiments, preferring the convenience and control offered by laboratory and glasshouse studies. Typically, when field studies have been conducted, they have emphasized the individual in preference to addressing ecosystem level questions. Another major limitation is technical. We

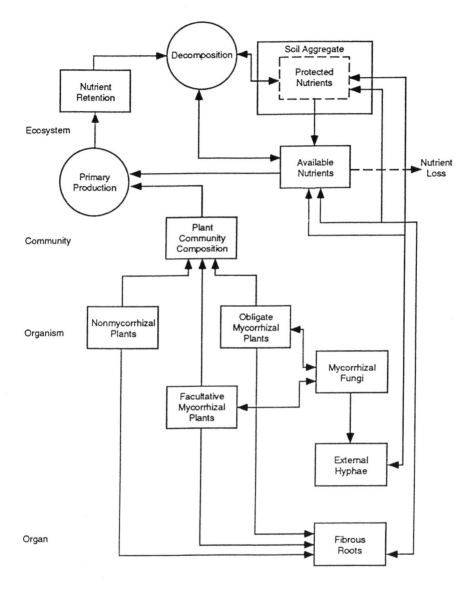

Figure 2. The linkage and feedbacks between mycorrhizae and biogeochemical cycling is hierarchically organized. The organizational level being illustrated is labeled on the left side of the diagram. Ecosystem processes are depicted by circles and measurable system components by blocks.

do not yet have the ability to easily quantify the biomass of most components of the VAM association.

Hence, to fully comprehend the contributions of VAM to biogeochemical cycling will require quantification of both vegetation and mycorrhizal fungi at temporal and spatial scales relevant to ecosystem studies. Unfortunately, our inability to do this is currently a major obstacle to including mycorrhizae in the mass balance of elements in terrestrial ecosystems.

The aim of this chapter is to identify those points where VAM fungi exert an influence on nutrient and biogeochemical cycling. First, we emphasize the contributions of VAM fungi, particularly external hyphae, by discussing their direct role in nutrient uptake and their indirect role in the accrual and mineralization of organic matter resulting from the formation of soil aggregates. Then we touch briefly on how the host response to the mycorrhizal association contributes to biogeochemical cycling. Finally, we discuss an example of how the interrelationship of mycorrhizae and community structure can influence biogeochemical cycling.

CONTRIBUTIONS OF MYCORRHIZAL FUNGI

The external hyphal network of VAM fungi plays an important role in nutrient uptake, especially for those ions that are not very mobile in soil solution. The classical explanation for the mechanism behind a mycorrhizally mediated increase in ion uptake is that the external hyphal network explores soil beyond the root hair zone. It is now becoming apparent, however, that the underlying mechanism for hyphal uptake of nutrients is more than a surface area phenomena (32).

Mycorrhizal hyphae may also foster a more integrated nutrient cycle. The occurrence of hyphal links between two or more plants is a common phenomenon in grassland and forest floor communities (38). Mineral nutrients can be passed between two plants via these hyphal links. Nutrients can also be passed from dying roots directly to living roots via hyphal connections (39). Newman (38) suggested that

mycorrhizal hyphal links promote direct nutrient
cycling by avoiding the mineralization process.
Hence, keeping more nutrient ions in biomass pools
increases the system's productivity.

Beyond their involvement in nutrient uptake, VAM
fungal hyphae are also believed to play an important
role in the physical enmeshment of soil particles to
create relatively stable soil aggregates (36). Thus,
VAM fungi indirectly influence the residence time of
nutrients within the system by promoting soil
aggregate development, which, in turn, reduces the
loss of soil (and nutrients) via erosion (35,36).
Moreover, the creation of an aggregated soil system is
part of the mechanism by which soils accrue organic
matter and organically bound nutrients (58).

Hence, the hyphae of VAM fungi are involved with
the scavenging and retention of nutrient ions and with
the creation of an aggregate system that acts as a
control point for accrual and mineralization of
organic matter in soils. At a larger scale, the
mycorrhizal association, by its involvement in
nutrient accumulation and retention, creates a system
that reduces erosion and leaching loss of nutrients
and trace ions. These issues are elaborated next.

The External Phase

Of the diverse kinds of organisms found in soil,
it is the mycorrhizal fungus, through its external
hyphal structure, that provides a direct physical link
between vegetation and the soil resource. Our
characterizations of the mycorrhizal association are
usually based on the temporal and spatial patterns of
colonized mycorrhizal roots (48,49). Few studies have
integrated measurements of colonized roots, external
hyphae, and vegetation (34,37). Very little
information is available regarding the amounts of
external hyphae in soils because this phase is
difficult to quantify (11,47). In most studies that
have attempted to quantify external hyphae, the host
plants were grown in containers under glasshouse
conditions. Of the relatively few field studies in
which external hyphae were quantified, most involved
sandy soils and dune systems. The techniques for

extracting and quantifying VAM fungal hyphae from heavy soils have only recently been developed (2,37).

A wide range of external hyphal lengths has been reported for field studies (Table 2). The reported differences are probably related to ecosystem properties such as vegetation, soil type, and rooting density. Unfortunately, most of these studies give little information on the hyphal extraction efficiency and counting accuracy of their procedures, so correlations of hyphal abundance with causative factors are problematic. Furthermore, there is a broader issue concerning the identification of "VAM hyphae." For example, some studies use hyphal diameter to separate VAM fungal hyphae from saprophytic fungal hyphae (5,53). Other studies have based their identification on comparative morphology of the hyphae (2,37,40). The resolution of this particular problem is not likely until the molecular techniques currently being developed to identify VAM fungi are refined and become available for routine application.

The factors that best predict external hyphal length in soils are root morphology and organic carbon content. For temperate grassland ecosystems, coarser fibrous root systems have more external VAM fungal hyphae associated with them than do finer fibrous root systems (34,37). In addition, plant life form appears to be a good predictor of root fibrosity and hence of external hyphae (34). In tropical grasslands a positive association exists between external hyphal length and soil organic matter over a range of soil textural classes (33).

The contributions of external VAM fungal hyphae to biogeochemical cycling are fivefold. First, external hyphae compose a rather large carbon sink, and hence their turnover may represent a significant control point for carbon and nutrient cycling within the rhizosphere. Second, external hyphae can chemically alter the soil matrix around their hyphal network, thereby increasing the availability of mineral ions to plant and microbial biomass. Third, the physical placement of external hyphae within the soil matrix allows access to nutrient ions that are beyond the root and root hair zone.   Fourth, hyphal links between roots of two or more plants allow for the potential transfer of nutrients directly from one root system to another. Last, mycorrhizal fungal hyphae contribute to the formation of stable soil aggregates, a necessary process for the accrual of organic matter and nutrients in soil.   The first four mechanisms suggest that VAM fungal hyphae facilitate "tighter" nutrient cycling by keeping nutrient ions in plant biomass rather than allowing their loss to the mineral ion pool.   Through the fifth mechanism, VAM fungal hyphae influence the storage of nutrients in organic pools and the rate of mineralization from these pools.   The first four mechanisms are discussed below, and the fifth is elaborated in the next section.

Recent studies of belowground carbon allocation using ^{14}C have determined that external VAM fungal hyphae represent approximately 26% of the labeled extraradical organic carbon pool (20).   This finding suggests that VAM fungal hyphae may be an important source of carbon to soil microorganisms.   Because hyphal cell walls also contain chitin, they may also be a source of nitrogen, albeit not a readily available one.   The availability of carbon and

Table 2.   The amounts of external hyphae for **VAM** fungi from various communities and soil types.

Community	Soil type[a]	External hyphae	Reference
Shrubland	scl	2.6 km kg^{-1}	4
Grassland	cl	54 km kg^{-1}	2
Dunes	s	12 m g^{-1}	53
Early-fixed dune	s	4.5 g l^{-1}	40
Virgin grassland	sl	19 m g^{-1}	56
Tallgrass prairie	sicl	45 m cm^{-3}	37
Corn field	sicl	17 m cm^{-3}	37

[a] scl = sandy clay loam; cl = clay loam; s = sand; sl = sandy loam; sicl = silt clay loam

nutrients derived from hyphal biomass is dependent
upon their turnover rates.

Very little information is available on either the
amount of external hyphal production or hyphal
turnover rates. St. John and Coleman (50), using the
data of Nicolson and Johnston (40), calculated a
through-put production estimate for external VAM
fungal hyphae of 120-400 g m^{-2} in a maritime dune
system. In a chronosequence of tallgrass prairie
reconstructions, Miller and Jastrow (37) observed an
increase in the exponential growth rate for hyphae of
39% for each growing season to a maximum yield of 47 m
cm^{-3}. Another study at this same site (Miller,
Reinhardt, and Jastrow, in preparation) estimates the
annual production of external hyphae to be 28 m cm^{-3},
with a calculated annual hyphal turnover of 26%.
However, because of an eight-week time window between
sampling dates, this value is most likely a low
estimate.

It has been suggested that turnover rates are
faster for VAM fungal external hyphae than for hyphae
within roots (17). This may be true for the
relatively thin-walled, small-diameter hyphae, but a
substantial portion of the VAM hyphal network is
composed of thick-walled runner or arterial hyphae
(14,47). These runner hyphae are likely to be longer
lived and more recalcitrant than the thinner walled
hyphae that are probably directly involved in nutrient
acquisition. Furthermore, a considerable proportion
of the external hyphae extracted from soil is either
nonviable or highly vacuolated (17,52,54), suggesting
considerable persistence of these hyphae.

The external hyphae of mycorrhizal fungi can also
influence nutrient cycling within the rhizosphere by
influencing soil pH and by producing organic acids.
The usual explanation for the elevated P levels found
in shoots of mycorrhizal plants is the increased
surface area provided by the external hyphae
associated with their root systems. However, it now
appears that this view is somewhat simplistic. Like
roots, hyphae apparently can form zones of P depletion
and altered pH in the surrounding soil (32).
Acidification of the surrounding soil is believed to

be a mechanism for enhancing the mobility of Ca-bound phosphates and possibly trace nutrients. In addition to proton extrusion, the production of low-molecular-weight organic acids (e.g. oxalate) by external hyphae may be a mechanism behind soil acidification and increased P uptake (28,32). The acidification process may also be a factor in the weathering of micas, releasing K and Fe in the rhizosphere (31). In addition, soil acidification caused by elevated levels of CO_2 produced by mycorrhizal roots and hyphae may result in better utilization of soil P sources by a host (20,29).

Although the amount of external hyphae is probably an important factor in nutrient ion uptake, the positioning of the hyphae around the root and into the soil may be equally important. For example, just as root architecture influences nutrient scavenging (12), the architecture of the external hyphae may also affect nutrient uptake (47). Furthermore, the reported differences in P uptake among the isolates of VAM fungi were independent of the amount of external hyphae produced (19). Rather, the differences in P uptake appear to be related to the placement of hyphae in the surrounding soil (1, 19).

The transfer of nutrients via hyphal links could be an important contribution of mycorrhizae to biogeochemical cycling (21,38). These links may function to retain nutrients within biomass. For example, mycorrhizal links between dying and living roots could allow for a tighter cycling of nutrients (mainly P) because a rapid transfer of P from dying roots to roots of a living plant can occur if the two root systems are intermingled (38). Such a mechanism would be especially important in perennial ecosystems, where the turnover of roots represents a considerable portion of the available nutrient pool. These links would also allow the plant to access nutrients directly, without having to rely on the mineralization process.

Soil Aggregation

Although primary production is the main source of organic matter in soils, for many soils it is the creation of aggregates that facilitates the accrual of organic matter. Organic matter accumulates because organic residues are physically protected from the actions of soil organisms when they are encapsulated by clays and silts during the aggregation process (9, 58). The importance of soil aggregate structure to carbon and nutrient storage was demonstrated recently for mollisols (8,16). In these soils a considerable amount of nutrients exist in the "protected" aggregated state (Table 3).

The development of stable soil aggregates within a system also allows the creation of a nutrient reserve (35). The sources of a soil's nutrient reserve include mineral ions from atmospheric deposition and weathering, plus organically bound nutrients in detrital materials. Unless detrital inputs are stabilized, accumulations of organic matter and the concomitant buildup of a nutrient reserve are usually minimal.

Organic residues are stabilized or protected within soil aggregates by both physical and chemical binding mechanisms (10,41,58). Visual evidence suggests that saprophytic and VAM fungal hyphae are

Table 3. Effects of aggregation on N mineralization (μg g^{-1}) for macroaggregates and microaggregates from two different mollisols.

Aggregate size	N mineralization	
	0-14 days[a]	0-20 days[b]
Native soil		
Macroaggregates	44.6	68
Crushed macroaggregates	64.1	94
Microaggregates	21.4	44
Cultivated soil		
Macroaggregates	23.4	31
Crushed macroaggregates	27.5	42
Microaggregates	15.1	17

[a] Gupta and Germida (16).
[b] Elliott (8).

involved in both the physical and chemical binding of
soil aggregates (13,16,42). Additional evidence
supporting the role of mycorrhizal fungal hyphae in
physical and chemical binding were presented by
Tisdall (55) and Miller and Jastrow (36). Briefly, it
appears that a simple hyphal entanglement mechanism
apparently contributes to the formation of soil
aggregates during an initial "aggregative phase." Once
aggregates are formed, cementation of fungal hyphae to
soil particles by organic or amorphous materials is a
mechanism apparently involved in a later stabilization
phase (16).

The contribution of VAM fungal hyphae to the
formation of stable macroaggregates (> 0.25 mm
diameter) involves three closely related processes
(35,36). The first process involves the growth of
hyphae into the soil matrix to create the skeletal
structure that holds the primary soil particles
together via physical entanglement. A second process
is the creation, by roots and external hyphae, of the
conditions that are conducive to the formation of
microaggregates (< 0.25 mm diameter). In the third
process, microaggregates and smaller macroaggregates
are enmeshed by external hyphae and roots to create
macroaggregate structures. All three of the processes
can occur simultaneously.

Some soils may develop only slowly beyond the
first process, especially on sites where clay content
and organic matter are very low (e.g. on mine spoil
materials or on shoreline dunes). For soils that do
progress past the first process, microaggregates may
be formed within the centers of stable macroaggregates
(41). In addition, while macroaggregates are being
created, others are being degraded because of a
variety of factors including root growth,
decomposition of labile organic binding agents, or
local reductions in the growth of mycorrhizal fungi
(57). In contrast, microaggregates, once stabilized,
are only slowly degraded because of the greater
physical protection and chemical resistance of the
organic components (57).

The growth and placement of external VAM fungal
hyphae within the rhizosphere can probably be affected

by the amount and quality of organic debris in the soil. For example, some evidence suggests that external hyphae can respond to organic rich microsites by proliferating into them (51). The ability of VAM fungal hyphae to respond to microsite differences has been suggested as an underlying mechanism behind the physical entanglement of soil particles and the enmeshment of smaller aggregates to create larger ones (36). The occurrence of such organic rich microsites appears to be related to the activities and by-products of soil fauna (6,36).

The external hyphae of VAM fungi play an important role in stabilizing organic matter in soils because of their involvement in the aggregation process. Furthermore, evidence suggests that the aggregation process is a control point for microbial accessibility to organic matter and, thus, for nutrient mineralization (58). In addition, highly aggregated soils are likely to have a more diverse biotic community and food web structure, which probably is not only more resistant to perturbation but also facilitates tighter nutrient cycling. Hence, by contributing to the aggregation process, VAM fungi are involved in a feedback mechanism whereby they are both creating and responding to their immediate environment because of the influence of the aggregation process on nutrient storage and mineralization. Mycorrhizally mediated changes in soil aggregates also contribute to soil structural aspects of the ecosystem, which can affect nutrient cycling through physical properties such as infiltration, aeration, erosion, runoff, and hydraulic conductance (25).

CONTRIBUTIONS OF MYCORRHIZAL PLANTS

Of the many processes influenced by the VAM association (Table 1), some include such host functions as photosynthesis, nutrient uptake, and water usage. VAM influences on host functions can affect nutrient accumulation and can alter nutrient ratios in plant tissues. In addition, VAM can affect a host by altering ion concentration ratios (especially for P, Zn, and Cu) in host biomass

relative to the soil. VAM fungi can influence plant
nutrient content by affecting a host's growth rate and
by influencing the rate of mineral ion uptake
(7,23,30). All of these responses are important to
biogeochemical cycling because nutrients residing in
biomass are not as mobile as those found in the soil
solution. Nutrients tied up in biomass are,
therefore, less likely to be lost from the system. As
a caveat to this mechanism, much of what we know about
mycorrhizal influences on nutrient uptake and
concentration ratios comes from plants grown in
containers. At present, reports from field studies of
VAM effects on plant growth and nutrient accumulation
are conflicting (12). A consequence of a VAM-mediated
increase in nutrient ion accumulation is that
nutrients immobilized in plant and microbial biomass
may be more readily accessible via mineralization than
nutrients residing in mineral forms that still require
geochemical processes (i.e. weathering) for their
release. If the process of nutrient accumulation in
biomass is extended over a larger time scale and a
soil accumulates organic matter well beyond simple
maintenance levels (to a high level of carbon
storage), the saprophytic process of mineralization
may replace ion scavenging by VAM fungal hyphae as the
control mechanism limiting plant nutrient uptake.
Although no evidence currently exists to support this
scenario, studies of succession in mesic tallgrass
prairie ecosystems suggest that after an initial
dependence on weathering processes, a greater reliance
on mycorrhizae will occur during midsuccession than in
late succession (Miller and Jastrow, unpublished
data). In late successional systems or those where
sufficient organic matter has accumulated, the
importance of mycorrhizae to plants may be in the
competition for mineralized nutrients and in the
effective closing of nutrient cycles (21,22).

MYCORRHIZAE AT HIGHER LEVELS OF ORGANIZATION

The relationships of plant species composition and
soil nutrient quality to biogeochemical cycling are
usually viewed as direct. However, this view of the

community ignores a major control point for community interactions, the mycorrhizal association. Mycorrhizae appear to mediate many community level processes and phenomenon. Table 1 shows the floristic properties of diversity and stability as being influenced by mycorrhizal fungi at the community level of organization. Grime et al. (15) conducted a microcosm study that strongly suggests that mycorrhizal fungi play an important role in maintaining floristic diversity. The mechanism facilitating the maintenance of diversity appears to be a mycorrhizally mediated export of assimilate from canopy dominants to subordinate species through common hyphal threads. As previously discussed, hyphal links between neighboring hosts can also foster direct nutrient transfer, thereby, reducing the dependence of a smaller cohort on the direct uptake of mineralized nutrients (38).

There is also a strong feedback between the species composition of vegetation and nutrient cycling. Research with prairie grasses indicates that individual plant species can affect mineralization and nutrient availability in their rhizosphere which, in return, lead to feedbacks between processes controlling species composition and ecosystem processes (59). The mechanisms driving these feedbacks include changes in belowground tissue N concentrations, belowground lignin concentrations, and belowground biomasses of the plant community caused by changes in vegetative composition. Furthermore, changes in plant species composition and nutrient cycling can act as an important selective force on VAM fungal populations, either directly through changes in internal root environments or indirectly through influences on surrounding soil conditions (27).

In tallgrass prairie ecosystems, the warm-season grasses are able to produce biomass with low investments of N and are able to allocate most of their biomass belowground (59). Observations that these plants are able to deplete the amount of available N to low levels, indicates they are good competitors under N-limiting conditions. Also, the low quality of their litter creates an environment

where the cycling of mineral ions and N-supply rates
are slow. These conditions also favor the need for
mycorrhizae as a nutrient scavenging strategy. In
contrast, many of the co-occurring cool-season
Eurasian grasses allocate less biomass belowground and
appear to have a greater demand for N (59). In
addition, they tend to have higher tissue N
concentrations, and their litter decomposes more
rapidly. The accumulation of litter and biomass with
high N content creates an environment that favors a
higher nutrient supply, and hence, these plants are
not as dependent on mycorrhizally supplied nutrients
as are warm-season prairie grasses.

It has also been demonstrated that many
warm-season prairie grasses are obligate in their need
for mycorrhizally supplied nutrients, whereas, most
cool-season grasses appear either to have a
facultative dependence or are nondependent on
mycorrhizae (18). A major contributing factor to
mycorrhizal dependency is the morphology of the root
system (18), with warm-season grasses producing coarse
fibrous roots and cool-season grasses producing fine
fibrous roots.

Consequently, within the tallgrass prairie those
plants that have the greatest need for mycorrhizally
supplied nutrients also create a soil and litter
environment with a low nutrient supply rate (although
stored nutrient potential may be quite high). The
mycorrhizal association, thus, contributes to the
creation of a positive feedback mechanism between
nutrient cycling and the maintenance of species
composition.

Changes in plant species composition can also
affect biogeochemical cycling by affecting the soil
aggregation process (24). Miller and Jastrow (34)
have hypothesized that observed associations between
plant species composition and aggregate development
were related to the types of roots produced by the
various species. They found that a large portion of
the effects of roots on soil aggregation was due to
indirect effects of root associations with mycorrhizal
fungi. Because of the effects of plant lifeform on
root morphology, some lifeforms may be more effective

than others in promoting aggregate formation. Hence, one of the mechanisms behind plant community composition acting as a control point for organic matter accumulation (nutrient cycling) is based on the relationship between the plant's root morphology and mycorrhizal fungi.

CONCLUSIONS

Mycorrhizae form a fundamental link between the biotic and geochemical portions of the ecosystem (43). A major area of neglect in mycorrhizal research is the integration of biogeochemical cycling with plant and mycorrhizal fungal responses to their biological and physicochemical environment. The primary reason why this area has been neglected is that these topics of research are usually investigated at different scales. Typically, the interactions of plants and VAM fungi with soils are studied at lower levels of organization within the ecosystem (e.g. the individual) than is biogeochemical cycling. The major obstacles for understanding the role of mycorrhizae in biogeochemical cycling are both conceptual and technical. Investigations of the contributions of mycorrhizae to biogeochemical cycling will require the quantifying of both plants and mycorrhizal fungi and measurements of their function at both temporal and spatial scales relevant to ecosystem studies of nutrient ion storage and flux.

We have argued in this paper that mycorrhizae contribute to biogeochemical cycling in more ways than just as nutrient scavengers. Over large time scales these contributions can be summarized as follows: (a) mycorrhizae influence the residence time of nutrients by altering nutrient concentration ratios in the vegetation; (b) mycorrhizae decrease the mobility of nutrients by placing a greater proportion of nutrient ions in biomass (i.e., the system becomes less leaky); and (c) as a consequence of the former two contributions, mycorrhizae increase the reliance of the system on mineralization processes rather than on soil parent material pools for acquiring nutrients. The first two processes are a direct consequence of interactions between the various plant species in the community and the associated VAM fungal hyphae. The third process is the result of a buildup of organic matter in the system and requires the protection of organic matter via the soil aggregation process. Although the literature on the contributions of VAM fungal hyphae to biogeochemical cycling is virtually nonexistent, mycorrhizal contributions are substantial and are important to biogeochemical cycling.

LITERATURE CITED

1.   Abbott, L. K., and Robson, A. D.   1985. Formation of external hyphae in soil by four species of vesicular-arbuscular mycorrhizal fungi.   New Phytol. 99:245-255.

2.   Allen, E. B., and Allen, M. F.   1986.   Water relations of xeric grasses in the field: Interactions of mycorrhizas and competition. New Phytol. 104:559-571.

3.   Allen, M. F., Clouse, S. D., Weinbaum, B. S., Jeakins, S. L., Friese C. F., and Allen, E. B. 1992. Mycorrhizae and the integration of scales: From molecule to ecosystem. Pages 488-515 in: Mycorrhizal Functioning. M. F. Allen, ed., Chapman and Hall, Inc. New York, NY.

4.   Allen, M. F., and MacMahon, J. A.   1985.   Impact of disturbance on cold desert fungi: Comparative microscale dispersion patterns. Pedobiologia 28:215-224.

5.   Ames, R. N., and Bethlenfalvay, G. J.   1987. Mycorrhizal fungi and the integration of plant and soil nutrient dynamics.   J. Plant Nutrition 10:1313-1321.

6.   Anderson, J. M.   1988.   Spatiotemporal effects of invertebrates on soil processes.   Biol. Fert. Soils 6:216-227.

7.   Barea, J. M.   1991.   Vesicular-arbuscular mycorrhizae as modifiers of soil fertility. Adv. Soil Science 15:1-40.

8.   Elliott, E. T. 1986. Aggregate structure and carbon, nitrogen, and phosphorus in native and cultivated soils. Soil Sci. Soc. Amer. J.

50:627-633.

9.   Elliott, E. T., and Coleman, D. C. 1988. Let the soil work for us. Ecol. Bull. 39:23-32.

10. Emerson, W. W., Foster, R. C., and Oades, J. M 1986. Organo-mineral complexes in relation to soil aggregation and structure. Pages 521-548 in: Interactions of Soil Minerals with Natural Organics and Microbes. P. M. Huang and M. Schnitzer, eds., SSSA Special Publ. No. 17. Soil Science Society of America, Madison, WI.

11. Finlay, R., and Söderström, B. 1992. Mycorrhiza and carbon flow to the soil. Pages 134-160 in: Mycorrhiza Functioning. M. F. Allen, ed., Chapman and Hall, New York, NY.

12. Fitter, A. H. 1985. Functioning of vesicular-arbuscular mycorrhizas under field conditions. New Phytol. 99: 257-265.

13. Foster, R. C. 1985. *In situ* localization of organic matter in soils. Quaestiones Entomologicae 21:609-633.

14. Friese, C. F., and Allen, M. F. 1991. The spread of VA mycorrhizal fungal hyphae in the soil: Inoculum types and external hyphal architecture. Mycologia 83:409-418.

15. Grime, J. P., Mackey, J. M. L., Hillier, S. H., and Read, D. J. 1987. Floristic diversity in a model system using experimental microcosms. Nature 328: 420-422.

16. Gupta, V. V. S. R., and Germida, J. J. 1988. Distribution of microbial biomass and its activity in different soil aggregate size classes as affected by cultivation. Soil Biol. Biochem. 20:777-786.

17. Hamel, C., Fyles, H., and Smith, D. L. 1990. Measurement of development of endomycorrhizal mycelium using three different vital stains. New Phytol. 115:297-302.

18. Hetrick, B. A. D., D. G. Kitt, and Wilson, G. T. 1988. Mycorrhizal dependency and growth habit of warm-season and cool-season tallgrass prairie plants. Can. J. Bot. 66:1376-1380.

19. Jakobsen, I., Abbott, L. K., and Robson, A. D. 1992. External hyphae of vesicular-arbuscular

mycorrhizal fungi associated with *Trifolium subterraneum* L. Spread of hyphae and phosphorus inflow into roots. New Phytol. 120:371-380.

20. Jakobsen, I., and Rosendahl, L. 1990. Carbon flow into soil and external hyphae from roots of mycorrhizal cucumber plants. New Phytol. 115:77-83.

21. Janos, D. P. 1983. Tropical mycorrhizas, nutrient cycles and plant growth. Pages 327-345 in: Tropical Rain Forest: Ecology and Management. S. L. Sutton, T. C. Whitmore, and A. C. Chadwick, eds., Blackwell Scientific Publ., Oxford.

22. Janos, D. P. 1985. Mycorrhizal fungi: Agents or symptoms of tropical community composition? Pages 98-103 in: Proc. Sixth North American Conference on Mycorrhizae. R. Molina, ed., June 25-29, 1984. College of Forestry, Oregon State Univ., Corvallis, OR.

23. Jarrell, W. M., and Beverly, R. B. 1981. The dilution effect in plant nutrition studies. Adv. Agronomy 34:197-224.

24. Jastrow, J. D. 1987. Changes in soil aggregation associated with tallgrass prairie restoration. Amer. J. Bot. 74: 1656-1664.

25. Jastrow, J. D., and Miller, R. M. 1991. Methods for assessing the effects of biota on soil structure. Agric., Ecosystems Environ. 34:279-303.

26. Jenny, H. 1980. The soil resource. Springer-Verlag, New York.

27. Johnson, N. C., Tilman, D., and Wedin, D. 1992. Plant and soil controls on mycorrhizal fungal communities. Ecology 73: 2034-2042.

28. Jurinak, J. J., Dudley, L. M., Allen, M. F., and Knight, W. G. 1986. The role of calcium oxalate in the availability of phosphorus in soils of semiarid regions: A thermodynamic study. J. Soil Science 142:255-261.

29. Knight, W. G., Allen, M. F., Jurinak, J. J., and Dudley, L. M. 1989. Elevated carbon dioxide and solution phosphorus in soil with vesicular-arbuscular mycorrhizal western

wheatgrass. Soil Sci. Soc. Am. J. 53:1075-1082.

30.   Koide, R. T. 1991. Nutrient supply, nutrient demand and plant response to mycorrhizal infection. New Phytol. 117:365-386.

31.   Leyval, C., Laheurte, F., Belgy, G., and Berthelin, J. 1990. Weathering of micas in the rhizospheres of maize, pine and beech seedlings influenced by mycorrhizal and bacterial inoculation. Symbiosis 9:105-109.

32.   Li, X.-L., George, E., and Marschner, H. 1991. Phosphorus depletion and pH decrease at root-soil and hyphae-soil interfaces of VA mycorrhizal white clover fertilized with ammonium. New Phytol. 119:397-404.

33.   McNaughton, S. J., and Oesterheld, M. 1990. Extramatrical mycorrhizal abundance and grass nutrition in tropical grazing ecosystem, the Serengeti National Park, Tanzania. Oikos 59:92-96.

34.   Miller, R. M., and Jastrow, J. D. 1990. Hierarchy of root and mycorrhizal fungal interactions with soil aggregation. Soil Biol. Biochem. 22:579-584.

35.   Miller, R. M., and Jastrow, J. D. 1992. The application of VA mycorrhizae to ecosystem restoration and reclamation. Pages 438-467 in: Mycorrhizal Functioning. M. F. Allen, ed., Chapman and Hall, Inc., New York, NY.

36.   Miller, R. M., and Jastrow, J. D. 1992. The role of mycorrhizal fungi in soil conservation. Pages 29-44 in: Mycorrhizae in Sustainable Agriculture. G. J. Bethenfalvay and R. G. Linderman, eds., ASA Special Publ. no. 54. Agronomy Society of America, Crop Science Society of America and Soil Science Society of America, Madison, WI.

37.   Miller, R. M., and Jastrow, J. D. 1992. Extraradical hyphal development of vesicular-arbuscular mycorrhizal fungi in a chronosequence of prairie restorations. Pages 171-176 in: Mycorrhiza in Ecosystems. I. J. Alexander, A. H. Fitter, D. H. Lewis, and D. J.

Read, eds.,   CAB International Publ., Oxon, UK.

38.   Newman, E. I.   1988. Mycorrhizal links between plants: Their functioning and ecological significance. Adv. Ecol. Res. 18: 243-270.

39.   Newman, E. I., and Eason, W. R.   1989. Cycling of nutrients from dying roots to living plants, including the role of mycorrhizas. Pages 133-137 in: Ecology of Arable Lands. M. Charholm and L. Bergström, eds., Kluwer Academic Publishers.

40.   Nicolson, T. H., and Johnston, C.   1979. Mycorrhiza in the Gramineae. III. *Glomus fasciculatus* as the endophyte of pioneer grasses in a maritime sand dune. Trans. Br. Mycol. Soc. 72:261-268.

41.   Oades, J. M.   1984.   Soil organic matter and structural stability: Mechanisms and implications for management. Plant Soil 76:319-337.

42.   Oades, J. M., and Waters, A. G.   1991. Aggregate hierarchy in soils. Aust. J. Soil Res. 29:815-828.

43.   O'Neil, E. G., O'Neil, R. V., and Norby, R. J. 1991. Hierarchy theory as a guide to mycorrhizal research on large-scale problems. Envir. Pollution 73:271-284.

44.   O'Neil, R. V., and Waide, J. B.   1981. Ecosystem theory and the unexpected: Implications for environmental toxicology. Pages 43-73 in: Management of Toxic Substances in Our Ecosystems. B. W. Cornaby, ed., Ann Arbor Science, Ann Arbor, MI.

45.   Perry, D. A., Amaranthus, M. P., Borchers, J. G., Borchers, S. L., and Brainerd, R. E. Bootstrapping in ecosystems. Bioscience 39:230-237.

46.   Read, D. J.   1991. Mycorrhizas in ecosystems. Experientia 47: 376-391.

47.   Read, D. J. 1992.   The mycorrhizal mycelium. Pages 102-133 in: Mycorrhizal Functioning. M. F. Allen, ed., Chapman and Hall, New York, NY.

48.   Reinhardt, D. R., and Miller, R. M.   1990. Size class of root diameter and mycorrhizal colonization in two temperate grassland

communities. New Phytol. 116:129-136.

49. Sanders, I. R., and Fitter, A. H. 1992. The ecology and functioning of vesicular-arbuscular mycorrhizas in co-existing grassland species. I. Seasonal patterns of mycorrhizal occurrence and morphology. New Phytol. 120: 517-524.

50. St. John, T. V., and Coleman, D. C. 1983. The role of mycorrhizae in plant ecology. Can. J. Bot. 61:1005-1014.

51. St. John, T. V., Coleman, D. C., and Reid, C. P. P. 1983. Growth and spatial distribution of nutrient-absorbing organs: Selective exploitation of soil heterogeneity. Plant Soil 71: 487-493.

52. Schubert, A., Marzachi, C., Mazzitelli, M., Cravero, M. C., and Bonfante-Fasolo, P. 1987. Development of total and viable extraradical mycelium in the vesicular-arbuscular mycorrhizal fungus *Glomus clarum* Nicol. & Schenck. New Phytol. 107: 183-190.

53. Sylvia, D. M. 1986. Spatial and temporal distribution of vesicular-arbuscular mycorrhizal fungi associated with *Uniola paniculata* in Florida foredunes. Mycologia 78:728-734.

54. Sylvia, D. M. 1988. Activity of external hyphae of vesicular-arbuscular mycorrhizal fungi. Soil Biol. Biochem. 20:39-43.

55. Tisdall, J. M. 1991. Fungal hyphae and structural stability of soil. Aust. J. Soil Res. 29:729-743.

56. Tisdall, J. M., and Oades, J. M. 1980. The effect of crop rotation on aggregation in a red-brown earth. Aust. J. Soil Res. 18:423-433.

57. Tisdall, J. M., and Oades, J. M. 1982. Organic matter and water-stable aggregates in soils. J. Soil Science 33:141-163.

58. Van Veen, J. A., and Kuikman, P. J. 1990. Soil structural aspects of decomposition of organic matter by micro-organisms. Biogeochem. 11:213-233.

59. Wedin, D. A., and Tilman, D. 1990. Species effects on nitrogen cycling: A test with perennial grasses. Oecologia: 433-441.

EFFECTS OF ECTOMYCORRHIZAE ON BIOGEOCHEMISTRY AND SOIL STRUCTURE

Janet S. MacFall
Duke University Medical Center
Durham, North Carolina

The role of mycorrhizal associations, especially ectomycorrhizae, in nutrient mineralization and uptake has been the subject of speculation by researchers concerned with tree establishment and growth for many years. As their beneficial nature became recognized, reports regarding their role in plant nutrition suggested that ectomycorrhizae function in the selective uptake of a number of nutrients, including N, P, K, Al, and Fe. It has not been conclusively determined whether benefits derived from mycorrhizal associations result simply from increased absorptive surface area, or if accelerated nutrient mineralization within the mycorrhizosphere is altering nutrient flux. As work continues, however, it has become clear that ectomycorrhizae have the capability of altering the rhizosphere biogeochemistry, and of creating mineralization patterns which differ from those of bulk soil.

HISTORY

The role of mycorrhizae in host plant nutrition and health has been the subject of investigation and interest, often with opposing views, since the initial description of mycorrhizae (16). There has been significant debate regarding the importance of mycorrhizae in plant nutrition, with early researchers' views ranging from the opinion that the fungal associations were pathogenic and not symbiotic as expressed by Robert Hartig and his students, to the position that they were normal, widespread and of mutual benefit to the fungus and host plant, as supported by published reports of Frank and Muller (18). Hatch (18) noted that one of the most widely used German forestry texts during the 1800's, carried through 11 editions by three generations of the Hartig family, included drawings of the absorbing roots of fir, and which illustrated mycorrhizae. A network of intercellular mycelium within the short roots is shown and described in texts by Th. Hartig (father of R. Hartig), but the character and significance of the fungal association was not recognized. In the 1870's, drawings and descriptions of beech mycorrhizal roots growing between decaying leaves were published by Muller (39), showing roots that were tightly bound to decomposing leaves through the hyphal network. He held the view that the fungus provided intimate contact between the decaying leaves and the roots, facilitating nutrient transport from leaf decomposition, and supporting the concept of a mutually beneficial relationship (18).

By the early 1900's, the abundance and widespread nature of ectomycorrhizal associations on trees had become accepted, however, the role of these associations was still not understood. Surveys of the roots of many forest trees throughout Europe (G. Sarauw in Denmark, France and Germany, E. Stahl in Germany, E. Melin in Sweden), Japan, the tropics and North America showed that from the first year of growth, abundant mycorrhizae were found to be universal (18). Mycorrhizae had been observed in a wide range of environments including forest habitats, open fields, alpine and arctic regions, sandy soils, heavy soils, and generally anywhere appropriate tree roots were present.

Observations regarding the relationship between plants of different species, the soil nutrient environment (particularly nitrogen) and mycorrhizal associations led E. Stahl (45) to propose one of the most perceptive hypotheses relating mycorrhizal associations to plant nutrition. He proposed that the abundance of mycorrhizae increases with soil poverty in mineral salts, particularly, but not limited to, nitrates. Stahl believed that mycorrhizal associations are most likely to form in plants which are limited in nutrient uptake, either due to restricted root systems and/or slow rates of transpiration and nutrient transport to the shoot. He noted a high efficiency for nutrient extraction from organic sources by fungi compared to plants, and speculated that as forest soils are frequently deficient in nitrates, the mycorrhizal associations allow plants to compete with fungi for available nutrients released from decaying organic material.

Stahl's work led to the development by E. Melin (37) of "The organic nitrogen theory", which proposes that organic nitrogen can be extracted from humus more efficiently by mycorrhizal than nonmycorrhizal plant roots. He observed that in bog soils, where N mineralization is primarily from organic N sources and nitrification is negligible, seedling growth was proportional to mycorrhizal establishment on the roots. This concept was the focus of most of the mycorrhizal research in the early 1900's extending into the 1930's. Subsequent studies by many researchers, however, failed to find the relationship between soil N availability and mycorrhizal abundance postulated by this theory, and raised questions concerning the utilization of N by mycorrhizae as being the sole benefit to the plant.

Cytological studies showing the high degree of physical contact between the hyphae and host cells strongly suggested that ectomycorrhizae play a role in regulation of water and general nutrient uptake. From Hatch (18) it is stated "From anatomical relationships it is known that all water and nutrients which enter tree roots (that is, forming ectotrophic mycorrhizae) by way of short roots pass through fungal hyphae, but in what manner and to what extent this is of significance to the tree is largely conjectural." Despite many publications since Hatch's report describing mycorrhizal effects on plant nutrition, the role of these fungi in nutrient mineralization and transport is not fully understood.

Subsequent work by Hatch (18) showed a decrease in development of mycorrhizae with increasing soil

concentrations of P, K, Ca and N. Further studies
showed greater nutrient uptake by mycorrhizal than
nonmycorrhizal roots, but with no evidence for
increased uptake of N from organic sources. He
concluded that benefits from mycorrhizal associations
were entirely due to the increased surface area
resulting from multiple branches of the short roots
and the extramatrical hyphae.

Mycorrhizal research next focused on increased
uptake of other nutrients, especially phosphorus.
Autoradiographic studies showed ectomycorrhizal short
roots had greater ^{32}P uptake than nonmycorrhizal roots
(22,36), with translocation through the fungus to pine
roots, stems and needles (38). Not only were
mycorrhizal roots shown to take up greater amounts of
phosphorus than nonmycorrhizal roots, but different
mycorrhizal types varied in rate of uptake,
temperature sensitivity and seasonal patterns (23,
35). Phosphorus uptake was shown to be an active,
metabolically-mediated process (44). Uptake of ^{32}P has
been demonstrated with transport not only to the host
tissue, but also into the mycelial network, with
transport measured as far as 40 cm through mycelial
strands (13).

MINERALIZATION AND TRANSPORT

The mechanisms facilitating the increased
nutrient uptake observed with ectomycorrhizal
associations are not well understood. Fungal
interactions with the soil are complex and variable,
and dependent both upon the specific fungus/host
combination and environmental conditions. This
interdependence forms a triangle of interactions
analogous to the plant disease triangle, wherein the
fungus, host and environment each impact the other,
determining if mycorrhizal colonization will occur,
and what the benefits to host and fungus will be.
Environmental factors strongly affecting colonization
and symbiotic benefits include soil conditions such as
soil type, fertility, pH, moisture, temperature and
potential nutrient sources.

Detailed examinations of the fungus-mediated

biogeochemistry impacting nutrient mineralization, transport, water flux, and weathering processes within the rhizosphere are scarce for ectomycorrhizal associations. Postulated mechanisms include increased uptake of minerals due to increased absorptive surface area, increased efficiency of uptake per unit surface area, promotion of rhizosphere microbial populations capable of accelerating mineralization, enzyme activities such as phosphatase enabling access to otherwise unavailable nutrient pools, production of siderophores capable of chelating nutrients within the rhizosphere and making them preferentially available to the fungus or the mycorrhizae, exudation of organic acids such as oxalic acid which may accelerate mineral weathering, and increased mineralization from greater fluctuations in soil moisture resulting from water extraction by mycorrhizal fungal hyphae.

Much of the past work regarding growth enhancement with ectomycorrhizae has focused on P mobilization and uptake. Reports regarding growth enhancement by mycorrhizae have been conflicting, some indicating that growth increases stimulated by fungal associations can be simulated by fertilizer applications, while others clearly demonstrated failure in tree establishment if mycorrhizae were absent, regardless of fertilization. This conflict was shown by McComb and Griffith (34) when they attempted to establish plantings of northern white pine (*Pinus strobus* L.) and Douglas-fir (*Pseudotsuga menziesii* (Mirb.) Franco) on a prairie site. Mycorrhizal inoculum was lacking in the soil and trees failed to establish mycorrhizae without artificial inoculations. White pines showed no difference in growth or tissue P concentration between trees receiving P fertilization or artificial inoculations. Douglas-fir, however, made satisfactory growth only with the introduction of mycorrhizal fungal inoculum, despite fertilizer applications.

Many other researchers have observed a similar inability to establish tree plantations in previously unforested sites. As summarized by Hatch (18), these regions included Western Australia, Southern Rhodesia, Java, Sumatra, the Philippines, and drained peat bogs in Sweden. More recently, *Pinus caribaea* Morelet establishment in Puerto Rico was successful only with mycorrhizal inoculations (48). Work by other groups has also demonstrated a more complex relationship, showing the degree of growth enhancement at varied phosphorus levels differed significantly between the fungi tested, and based on comparisons of relative root uptake efficiency, growth enhancement could not be due to P fertility alone (15,47). In contrast, the growth response of Lodgepole pine (*Pinus contorta* Doug. ex Loud) to fungal colonization could be simulated by fertilizer applications (41).

SPECIFIC EFFECTS ON MINERALIZATION, UPTAKE AND GROWTH

The role of an ectomycorrhizal fungus, *Hebeloma arenosa* Burdsall, MacFall & Albers, in nutrient mobilization, transport, and partitioning in a low fertility soil has been examined in some detail. This fungus has been found associated with a number of tree species in the Great Lake states, and appears to be tolerant to the high soil fertility conditions present in the nursery beds at these sites. It has been shown to be an effective soil inoculate for red pines (*Pinus resinosa* Ait.), increasing growth both of bareroot nursery and container-grown stock (28,29). This fungus also has been demonstrated to increase seedling survival following outplanting of container-grown seedlings to low fertility, sandy, droughty sites (29). To examine mycorrhizae/soil interactions, a low fertility sandy loam was collected from an area adjacent to the Wilson State Nursery, Boscobel, Wisconsin. Soil was amended with varied amounts of P as superphosphate (0-136 mg/kg). Half of each fertilizer treatment also received inoculum of *H. arenosa*, the other half received non-fungal carrier. Red pine seedlings were grown in the soil mixes for 19 weeks. During that period, seedlings were fertilized bi-weekly with 60 mg/kg N and K as ammonium sulfate and potassium sulfate, respectively. Experimental details and soil analyses are reported elsewhere (26). At harvest, the non-inoculated seedlings were seen to be nonmycorrhizal, while inoculated seedlings had formed abundant mycorrhizae when grown in P unamended soil. Fungal colonization decreased with increasing P additions to soil, so that at the highest level of P amendment, inoculated seedlings were also nonmycorrhizal. All of the nonmycorrhizal seedlings grown in the P unamended soil showed severe P deficiency symptoms, while none of the mycorrhizal seedlings grown in similarly P amended soil were symptomatic. At the mid-level of P amendment, 30% of the nonmycorrhizal seedlings exhibited purple needle tips, and symptoms of nutrient deficiencies were absent on nonmycorrhizal seedlings grown at the highest level of P amendment (26).

Root and shoot growth were both changed with mycorrhizal colonization and P fertilization, but in different fashions. Shoot and root dry weights for mycorrhizal seedlings were greater than for nonmycorrhizal seedlings at all but the highest level of P amendment. Shoot dry weights of mycorrhizal seedlings grown in P unamended soil were the same as for nonmycorrhizal seedlings which had received 34 mg/kg P, indicating that the fungal association had made an additional 34 mg/kg P available to the plant for shoot growth. Shoot growth of both inoculated and noninoculated seedlings was further increased with greater P additions to soil, however, showing that for shoots, growth enhancement observed with mycorrhizal associations could be simulated by P fertilization (26).

In contrast, maximum root dry weights were achieved only in mycorrhizal plants grown at low to mid levels of P application (26). Decreasing dry weights were observed in mycorrhizal plants with P additions and increasing dry weights observed in nonmycorrhizal plants. Only at the highest level of P amendment were the root dry weights of inoculated and noninoculated plants equal. These results indicate that the maximum root/shoot ratio, a growth strategy most favorable for survival during periods of stress, will be achieved by mycorrhizal plants grown at low to moderate levels of soil fertility. In addition, these results clearly demonstrate that the greater root growth promoted by *H. arenosa* could not be simulated by P fertilization.

In these experiments (26), greater root and shoot concentrations of P were measured in mycorrhizal seedlings compared to nonmycorrhizal seedlings when grown in P unamended soil. At mid and high levels of P amendment, however, there was no difference in tissue concentration between mycorrhizal and nonmycorrhizal seedlings. One interpretation of this observation is that at low P levels in soil, mycorrhizal roots are more efficient than nonmycorrhizal roots at P uptake. In endomycorrhizal systems, the fungal hyphae have a greater affinity for P than the host roots (8), however, this has not been conclusively demonstrated for ectomycorrhizae. The greater P tissue concentration, however, suggests that the mycorrhizal root affinity for P in soil with low P availability is greater than for nonmycorrhizal roots.

One potential mechanism for increased P uptake efficiency is through sequestration of a portion of the pool of biological P as polyphosphate. Uptake of P from soil solution is through processes of active uptake against a P concentration gradient. In an examination of the P partitioning within a mycorrhizal root system, it has been demonstrated by *in vivo* NMR spectroscopy that the mycorrhizal root contains a significant amount of the total P as polyphosphate, sequestered within the fungal vacuole. Polyphosphate is accumulated even when plants are grown under severely P limited conditions, effectively reducing the concentration of vacuolar and cytoplasmic P (31). Although tissue analysis of mycorrhizal root systems shows a higher concentration of P than for nonmycorrhizal roots, when much of that total P is removed from the cytoplasmic and vacuolar P pool through polyphosphate synthesis, the concentration gradient against which phosphate is transported when taken up from soil solution is significantly reduced. This mechanism has the potential of increasing the efficiency of P uptake by mycorrhizal roots compared to nonmycorrhizal roots (31).

Another interaction observed in these experiments is an increased uptake of K concurrent with increased tissue concentrations of P, both as a consequence of

mycorrhizal associations and P fertilization (26).
Fertilizer was applied for the duration of the
experimental period supplying N and K at "luxury"
levels to the plants. Although K was not limited in
supply and is relatively mobile in sandy soils,
increased tissue concentrations were found in
mycorrhizal seedlings grown in P limited soil,
compared to nonmycorrhizal plants, and in seedlings
receiving P fertilization, particularly in the plant
roots. A linear relationship was observed between P
and K tissue concentrations, both in mycorrhizal and
nonmycorrhizal seedlings. The relationship was more
pronounced in roots than in shoots (Figure 1).

As K was supplied in non-limiting amounts to all
of the plants in these experiments, increased uptake
of K as reflected in increasing tissue concentrations
must have been through increased selective uptake per
unit surface area. A suite of other elements were
included in the tissue analysis, including Al, B, Ca,
Co, Cr, Cu, Fe, Mg, Na, Ni, and Pb. A correlation
between the tissue concentration of any of these
elements and P was not observed, indicating a
selective uptake of K correlated with P uptake.

These results also indicate that, through
improved P nutrition, mycorrhizal roots have the
ability for greater K uptake than nonmycorrhizal
roots. The effect may be simulated by P
fertilization, however, most forest stands do not
receive P fertilization and are likely to depend upon
the mycorrhizal associations for mediation of the P
supply and uptake. Greater influx and reduced efflux
of K by mycorrhizal roots compared to nonmycorrhizal
roots has been demonstrated (43). Results shown here
suggest that this effect is related to P nutrition,
possibly through altered phospholipid content and
membrane permeabilities as has been suggested for the
relationship between P nutrition and root exudates.

In addition to increased surface area,
mycorrhizae have demonstrated an ability to solubilize
organic P substrates through the action of acid
phosphatase (acid phosphomonoesterase) (EC 3.1.3.2).
Organically bound phosphate may constitute 50-67% of
the total soil phosphorus supply and cannot be

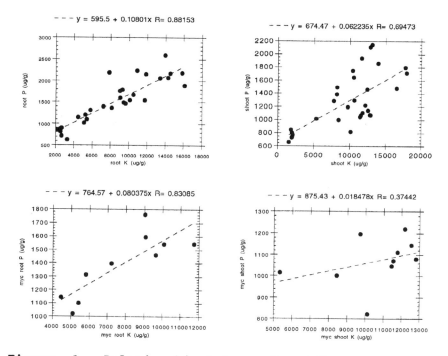

Figure 1. Relationship between P and K concentrations in roots and shoots of mycorrhizal and nonmycorrhizal plants. Graphs in the top row include tissue concentration data for all seedlings, grown at all three levels of P addition. Graphs in the lower row include data only for mycorrhizal seedlings grown in soil with no or mid levels of P amendment.

utilized by plants without the solubilization of orthophosphate. Acid phosphatase activity has been demonstrated in forest soil (19,40), and mycorrhizal fungi have been shown to have acid phosphatase activity *in vitro* (21), suggesting a role for mycorrhizal fungi in the solubilization of organic P sources. Several mycorrhizal fungi have been shown to utilize inositol hexaphosphate as a P source *in vitro* (5,46), presumably through enzymatic solubilization. In addition, excised mycorrhizae have demonstrated

acid phosphatase activity, with varied activity levels
between different species and isolates of
ectomycorrhizal fungi (3,4,20).

As indicated above, mycorrhizal red pines grown
in P deficient soil had greater dry weights and P
tissue concentrations than nonmycorrhizal seedlings
grown in the same soil (26). In addition to improved
P nutrition through increased surface area and
efficiency of uptake, increased solubilization of
organic P through enzyme activity has the potential of
increasing the pool of available P. To test this,
root segments of mycorrhizal and nonmycorrhizal
seedlings grown at varied levels of P amendment were
assayed with three different substrates for enzyme
activity (27). With all three assays, release of
orthophosphate into solution was measured as an
indicator of the ability to solubilize organic P
sources. One assay evaluated p-nitrophol release from
p-nitrophenol phosphate and was evaluated on a root
surface area basis, while the other two assays

**TABLE 1. Acid phosphatase activities of mycorrhizal
and nonmycorrhizal red pine seedlings grown in soil
with varied P amendments (27).**

| Treatment | umole P/mm2 | mg P/mg root | |
		Na Salt	KMg Salt
Low P			
Inoc.	.043d	36.4c	21.6c
Non-Myc.	.017a	0.0a	0.0a
Mid P			
Inoc.	.038c	51.3d	21.1c
Non-Myc.	.020b	32.1bc	2.3b
High P			
Inoc.	.081e	22.3b	1.7b
Non-Myc.	.081e	23.7b	2.8b

Numbers in the same column followed by different
letters were significantly different (Protected LSD,
p=0.05).

measured orthophosphate release into solution from two

myoinositol-hexaphosphate salts based on root dry
weights (27).

As the previous reports have demonstrated,
mycorrhizal roots have greater acid phosphatase
activity than nonmycorrhizal roots when grown in
similarly fertilized soil. In addition, each assay
(Table 1) showed a different change in enzyme activity
with respect to P fertilization, demonstrating
differing substrate affinities by the enzyme (27).
The p-nitrophenol phosphate assay showed enzyme
activity by all roots tested, with slightly decreased
activity in mycorrhizal roots grown in soil from
intermediate P applications, and the greatest enzyme
activity at the highest level of P addition. At the
highest level of fertilization, mycorrhizal
colonization was completely inhibited in the
inoculated plants, and all plants had actively
growing, white root tips. Actively growing roots have
been previously demonstrated to have high levels of
acid phosphatase activity when assayed with this
substrate, as was observed here. In contrast, when
assayed by P release from organic sources as would be
found in soil, nonmycorrhizal roots grown in P
unamended soil exhibited no detectable enzyme
activity. Roots of nonmycorrhizal plants also
exhibited reduced phosphate release when grown at the
highest levels of P addition. These results reflect,
as might be expected, variable affinities for
different substrates, and demonstrate a significant
difference in apparent patterns of enzyme activities
between the standard assay (p-nitrophenol phosphate)
and phosphate release from phytic acid salts.

ADDITIONAL BIOWEATHERING PROCESSES

Other compounds, such as organic acids and
siderophores, produced and released from mycorrhizae,
also are additional potential bioweathering agents.
Oxalic acid, when released into soil solution, has
been shown to solubilize Fe and Al, and to precipitate
with Ca (33). Calcium oxalate crystals have been
observed in the mantle of two very different
mycorrhizal types, suggesting the role for oxalic acid

in enhanced mineralization and effectively increasing the pool of nutrients available for extraction by ectomycorrhizal roots (32). In separate experiments, plant growth and uptake of P, Mg, Fe and K were greater for European beech (*Fagus silvatica* L.) inoculated with *Laccaria laccata* (Scop. ex Fr.) Cook than for nonmycorrhizal trees when grown in a growth medium containing only rock phosphate and mica as the sources for P, Fe, Mg and Al. The greater mineral dissolution was attributed to both increased root growth and production of greater amounts of organic acids within the mycorrhizosphere.

Production of siderophores by fungi has also been reported, with the speculation that siderophore production alters Fe availability within the rhizosphere. However, in comparison of slash pine (*Pinus elliottii* Engelm.) colonized by *Pisolithus tinctorius* (Pers.) Coker & Couch. and nonmycorrhizal plants, reduced Fe uptake was demonstrated with mycorrhizal associations for several chelates, including a partially purified siderophore produced by *P. tinctorius in vitro*. Prior growth in an Fe deficient medium increased Fe uptake by mycorrhizae and not by nonmycorrhizal roots, however, the mycorrhizal roots still had reduced uptake compared to roots of non-colonized plants (25). The differing rates of uptake were attributed to different mechanisms of Fe uptake between the fungus and roots.

WATER RELATIONS

In addition to increased mineralization and uptake of nutrients, ectomycorrhizal associations have an effect on host water relations. Transport of water from soil to plant via hyphal strands has been demonstrated (6), with variable hydraulic conductance between mycorrhizae (7). By use of Magnetic Resonance Imaging techniques, greater efficiency for water extraction from fine sand by mycorrhizal fine roots than by nonmycorrhizal taproots or lateral roots has been measured (30). Although taproots were seen to participate significantly in water uptake, ectomycorrhizal fine roots were clearly more efficient at water uptake. In these experiments, loblolly pine (*Pinus treda* L.) seedlings had been transplanted into fine sand, watered to field capacity, and repeatedly imaged. The water depletion zone could clearly be seen to form around specific roots over time. Due to the transplanting, the extensive mantle and hyphae penetrating into the surrounding soil had been removed. The greater extraction of water which was observed, therefore, was due to greater absorptive capabilities of the mycorrhizal fine roots themselves, without the extensive increase in surface area which would have been associated with extramatrical hyphae. As the water potential within the ectomycorrhizae cannot be less than the water potential of the lateral root for water to pass from the mycorrhiza into the transpirational stream, the greater uptake must be due to reduced resistance to water inflow into the mycorrhizal structure compared to the lateral root. This relationship can be clearly seen in Figure 2 and 3.

The greater water extraction by mycorrhizal than nonmycorrhizal short roots may have a significant impact on mineralization and transport processes of nutrients within the soil. For nutrients that are relatively mobile in soil, such as K and NH_4, greater nutrient uptake may be associated with greater water uptake as the nutrients travel with the water stream extracted by the mycorrhizae. In addition, fluctuating cycles of soil water within the mycorrhizosphere caused by the greater extraction of water by the mycorrhizae may accelerate mineralization, particularly N (49).

SIGNIFICANCE OF HYPHAL ORGANIZATION

Mycelial extensions also provide greater contact between the soil particles and the ectomycorrhizae, increasing the absorptive capacity not only by possessing greater surface area, but by increasing the physical contact with soil particles. Four general types of hyphal development have been described: rhizomorphs (*Rhizopogon*), the fairy ring (*Tricholoma*), irregular mats (*Hysterangium, Suillus*), and a diffuse

system (*Hebeloma, Amanita, Boletus, Russula*).
Multiple reports exist of transport of radiolabeled
materials between hosts and fungi, and the varied
forms suggest differences in uptake and transport
strategies (1). Rhizomorphs are considered
stress-resisting structures, resembling conductive
vascular cylinders, and are capable of transporting
water and materials. Fairy rings may range from
diffuse hyphae to fairly dense hyphal development in
soil. They frequently develop to encompass several
trees, extend with time, and may be quite old. The
mat occurs as a dense hyphal growth at the soil-litter
interface. Mats are usually composed of a dominant
single fungal species and may significantly alter the
nutrient cycling capabilities of a soil patch. The
diffuse hyphal system forms in an irregular pattern,
and may extend well into the soil.

    Fungal mats formed by *Hysterangium setchellii*
(Fisher) in association with Douglas-fir may colonize
up to 27% of the forest floor and account for 45-55%
of the total soil biomass (14). This fungus has been
demonstrated to immobilize large quantities of
nutrient in the fungal tissue (10), but also to
increase the rate of organic matter decomposition and
thus nutrient mineralization from organic sources (9).
Microbial biomass and rates of degradation of
cellulose (3-6 times) and lignin were significantly
higher in mat than in non-mat soils containing
ectomycorrhizae which had not formed mat structures
(11,12). There was not a significant, consistent
difference in moisture between mat and non-mat soils,
suggesting that the mats were not participating in
water extraction (17). Acetylene reduction,
reflective of nitrogen fixation, was also
significantly greater in the mat than in the non-mat
soils (17). In an additional study, it was suggested
that *H. setchellii*, through the formation of the
extensive hyphal mats (4x the biomass of adjacent
non-mat soils), increases litter decomposition,
resulting in more rapid release of N, P, K and Mg, and
more efficient removal of nutrients from soil solution
(12).

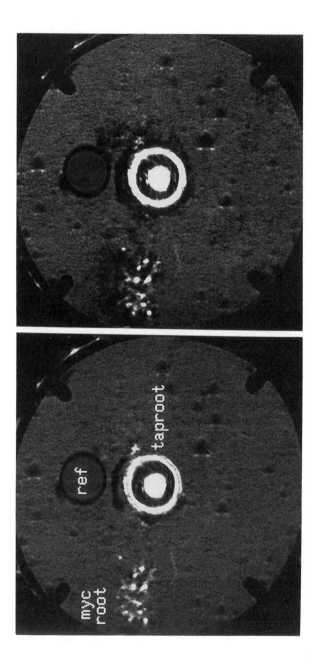

Figure 2. Magnetic resonance imaging of a cross-sectional view through a loblolly pine taproot and associated mycorrhizal root cluster (myc roots). The seedling has been planted into fine sand, watered to field capacity, then repeatedly imaged at the same slice position. On the left is a view at the beginning of the experiment, wherein the sand filling the potting container is uniformly bright. A reference tube (ref) is placed into the field of view to allow measurement of the water content of the sand directly from the acquired images. On the right is a view of the same seedling several hours later. Note the dark region surrounding both the taproot and the mycorrhizal root cluster. These dark regions are the water depletion zones formed by water extraction by the plant roots. The small diameter fine root attaching the mycorrhizal root cluster to the taproot can be faintly seen in the sand. Note that it is lacking a surrounding water depletion zone (MacFall, unpublished data).

Figure 3. Three-dimensional reconstruction of a
loblolly pine root system following a period of
transpiration. The seedling had been removed from the
growing container, root-pruned to leave only a single
lateral root and associated mycorrhizal fine roots,
replanted into a sand-filled container, watered to
field capacity, and studied with Magnetic Resonance
Imaging. The sand filling the container which was not
part of the water depletion zone has been removed from
the reconstruction, leaving only the water depletion
zone and associated root. The water depletion zone
can be seen as a "cloud" surrounding the taproot and
extending part-way down the lateral root. The
expansion of the water depletion zone near the end of
the lateral root was formed by water extraction by a
group of mycorrhizal short roots (MacFall, unpublished
data).

Even without the dense fungal mat formation, the diffuse hyphal network is successful both in accessing a significantly greater volume of soil for nutrient and water extraction, and for holding soil into micro- and macro-aggregates. A positive, significant correlation between biomass of *P. tinctorius* and P uptake by loblolly pine seedlings grown in sand culture has been observed, demonstrating a direct relationship between nutrient uptake and fungal growth (42).

The mycorrhizal type formed by *H. arenosa* is an extensive, wefty hyphal network, extending both down the lateral root and into the surrounding soil, creating an extensive mycorrhizosphere. This can be seen in Figure 4, wherein the soil held by the mycelial network is seen tightly clinging to the roots compared to the nonmycorrhizal seedling. This form of macro-soil aggregation may serve to stabilize the soil and to provide a significantly closer physical contact between the absorptive hyphae and the soil particles. In addition, the small diameter of fungal hyphae compared to roots, potentially allows them to enter smaller soil pores than could roots. The tendency for roots to follow paths of relatively less resistance, such as soil fractures and previously formed root channels, has the potential to reduce the physical connection between the root and the soil environment from which it must extract nutrients and water.

The close physical connection formed between soil particles and ectomycorrhizal hyphal extensions provides a spatially favorable environment for biological weathering processes and mineral uptake by the fungus. This close relationship has been suggested by SEM examinations of cross-sectional views of a fine root from a forest ecosystem embedded in resin-impregnated soil (2). Mineral grains were occasionally observed displaying an etch that replicated the root form, suggesting preferential mineral dissolution by the fine root. Although these observations are far from conclusive, they suggest a role for mycorrhizae in biologically accelerated mineralization processes.

Hebeloma   Low P   Control

Figure 4. Red pine seedlings grown in a sandy-loam. The seedling on the left is mycorrhizal with *Hebeloma arenosa*, an ectomycorrhizal symbiont common in the Great Lake states nurseries. Note the soil clumps held tightly to the root by the hyphal extensions from the mycorrhizae compared to the nonmycorrhizal seedling grown in the same soil, shown on the right.

Although by far most mycorrhizal associations are observed to form in the surface soil layers, mycorrhizae have been found much deeper in the soil profile. At a pit site in Duke Forest, North Carolina, mycorrhizae have been observed at depths > 1m, with hyphae adhering tightly to the clay fracture planes (D. Richter, pers. comm., J. MacFall, unpublished data). The role of these mycorrhizae in mineralization and nutrient acquisition is likely to be quite different from the physiological processes displayed by litter-colonizing mycorrhizae.

CONCLUSIONS

In summary, it can be concluded that ectomycorrhizae play a significant role in soil biogeochemistry and soil structure. Greater selective nutrient uptake from the increased absorptive surface area provided by the mycelial network has been shown with *H. arenosa* and other ectomycorrhizal associates. Although not clearly demonstrated, efficiency of uptake is likely to also be increased with mycorrhizae. Mechanisms for accelerated, biologically-mediated weathering of minerals and organic materials through the production of enzymes, organic acids, and siderophores are present in many ectomycorrhizae. Higher rates of carbon and nutrient mineralization have been observed within fungal mats compared to non-mat soils, suggesting the potential for a similar role in more diffuse hyphal structures. Significant water uptake and transport may also be accomplished through mycorrhizae, but is likely to differ between mycorrhizal types. Clearly these symbiotic associations have the potential to alter the soil chemistry of the mycorrhizosphere, and as our understanding of their physiological processes emerges, we can better develop a model for their role in nutrient mobilization and cycling at the ecosystem level.

LITERATURE CITED

1. Allen, M. 1992. The Ecology of Ectomycorrhizae. Cambridge Univ. Press, New York.
2. April, R., and Newton R. 1992. Minerology and mineral weathering. Pages 378-425 in: Atmospheric deposition and forest nutrient cycling. D. W. Johnson and S. E. Lindberg, eds., Springer-Verlag Ecological Studies 91. New York.
3. Antibus, R. K., Crosdale J. G., Miller O. K., and Linkins A. E. 1981. Ectomycorrhizal fungi of *Salix rotundifolia* III. Resynthesized mycorrhizal complexes and their surface phosphatase activities. Can. J. Bot. 59:2458-2465.
4. Bartlett, E. M., and Lewis D. H. 1973. Surface phosphatase activity of mycorrhizal roots of beech. Soil. Biol. Biochem. 5:249-257.
5. Bousquet, N., Mousain, D., and Salsac, L. 1985. Use of phytate by ectomycorrhizal fungi. Pages 363-368 in: Mycorrhizae: Physiology and genetics. V. Gianinazzi-Pearson and S. Gianinazzi, eds., Institut National de la recherge Agronomique, Paris. 831 pp.
6. Brownlee, C., Duddridge, J. A., Malibari, A., and Read, D. J. 1983. The structure and function of mycelial systems of ectomycorrhizal roots with special reference to their role in forming inter-plant connections and providing pathways for assimilate and water transport. Plant Soil 71:433-443.
7. Coleman, M. D., Bledsoe, C. S., and Smit, B. A. 1990. Root hydraulic conductivity and xylem sap levels of zeatin riboside and abscisic acid in ectomycorrhizal Douglas fir seedlings. New Phytol. 115:275-284.
8. Cress, W. A., Throneberry, G. O. and Lindsey, D. L. 1979. Kinetics of phosphorus absorption by mycorrhizal and nonmycorrhizal tomato roots. Plant Physiol. 64:484-487.
9. Cromack, K., Jr., Fichter, B. L., Moldenke, A. M., Entry, J. A., and Ingham, E. R. 1988. Interactions between soil animals and

ectomycorrhizal fungal mats. Agric., Ecosystems Environ. 24:161-168.

10. Entry, J. A., Stack, N. M., and Lowesenstein, H. 1987. Timber harvesting: effects on degradation of cellulose and lignin. For. Ecol. Manag. 22:79-88.

11. Entry, J. A., Donnelly, P. K., and Cromack, K. Jr. 1991. Influence of ectomycorrhizal mat soils on lignin and cellulose degradation. Biol. Fertil. Soils 11:75-78.

12. Entry, J. A., Rose, C. L. and Cromack, K. Jr. 1991. Litter decomposition and nutrient release in ectomycorrhizal mat soils of a Douglas-fir ecosystem. Soil. Biol. Biochem. 23:285-290.

13. Finley, R. D., and Read, D. J. 1986. The structure and function of the vegetative mycelium of ectomycorrhizal plants II. The uptake and distribution of phosphorus by mycelial strands interconnecting host plants. New Phytol. 103:157-165.

14. Fogel, R., and Hunt, G. 1979. Fungal and arboreal biomass in a western Oregon Douglas-fir ecosystem: Distribution and patterns of turnover. Can. J. For. Res. 9:245-256.

15. Ford, V. J., Torbert, J. L. Jr., Burger, J. A., and Miller, O. K. 1985. Comparative effects of four mycorrhizal fungi on loblolly pine seedlings growing in a greenhouse in a Piedmont soil. Plant Soil 83:215-221.

16. Frank, A. B. 1885. Uber die auf Wurzelsymbiose beruhende Ernahrung gewisserBaume durch unteriridische Pilze. Ber. d. deut. bot. ges 3:128-145.

17. Griffiths, R. P., Caldwell, B. A., Cromack, K., Jr., and Morita, R. Y. 1990. Douglas-fir forest soils colonized by ectomycorrhizal mats. I. Seasonal variation in nitrogen chemistry and nitrogen cycle transformation rates. Can. J. For. Res. 20:211-218.

18. Hatch, A. B. 1937. The physical basis of mycotrophy in pines. The Black Rock Forest Bull. no. 6., Dir., H. H. Tryon.

19. Ho, I. 1979. Acid phosphatase activity in forest

soil. For. Sci. 25:567-568.

20. Ho, I. 1987. Comparison of eight *Pisolithus tinctorius* isolates for growth rate, enzyme activity, and phytohormone production. Can. J. For. Res. 17:31-35.

21. Ho, I., and Zak, B. 1979. Acid phosphatase activity of six ectomycorrhizal fungi. Can. J. Bot. 57:1203-1205.

22. Kramer, P. J., and Wilbur, K. M. 1949. Absorption of radioactive phosphorus by mycorrhizal roots of pine. Science 110:8-9.

23. Langlois, C. G., and Fortin, J. A. 1984. Season variations in the uptake of (^{32}P) phosphate ions by excised ectomycorrhizae and lateral roots of *Abies balsamea*. Can. J. For. Res. 14:412-415.

24. Leyval, C., and Berthelin, J. 1989. Interactions between *Laccaria laccata*, *Agrobacterium radiobacter* and beech roots: Influence on P, K, Mg and Fe mobilization from minerals and plant growth. Plant Soil 117:103-110.

25. Leyval, C., and Reid, C. P. P. 1991. Utilization of microbial siderophores by mycorrhizal and nonmycorrhizal pine roots. New Phytol. 119:93-98.

26. MacFall, J. S., Slack, S. and Iyer, J. 1991. Growth of red pine (*Pinus resinosa* Ait.) as influenced by phosphorus applications and *Hebeloma arenosa*. Can. J. Bot. 69:372-379.

27. MacFall, J. S., Slack, S. A., and Iyer, J. 1991. Acid phosphatase activity of red pine roots as influenced by phosphorus applications and the ectomycorrhizal fungus *Hebeloma arenosa*. Can. J. Bot. 69:380-383.

28. MacFall, J. S., and Slack, S. A. 1991. Effects of *Hebeloma arenosa* on growth of red pine seedlings in high fertility nursery soil in Wisconsin. Can. J. For. Res. 21:482-488.

29. MacFall, J. S., and Slack, S. A. 1991. Effects of *Hebeloma arenosa* on growth and survival of container-grown red pine seedlings (*Pinus resinosa* Ait.) Can. J. For. Res. 21:1459-1465.

30. MacFall, J. S., Kramer, P. J., and Johnson, G. A.

1991 Comparative water uptake of pine seedling roots as determined by Magnetic Resonance Imaging. New Phytol. 119:551-560.

31. MacFall, J. S., Slack, S. A. and Wehrli, S. 1992. Phosphorus distribution in red pine roots and the ectomycorrhizal fungus *Hebeloma arenosa*. Plant Physiol. 100:713-717.

32. Malajczuk, N., and Cromack, K. Jr. 1982. Accumulation of calcium oxalate in the mantle of ectomycorrhizal roots of *Pinus radiata* and *Eucalyptus marginata*. New Phytol. 92:527-531.

33. Marshner, H. 1986. Mineral Nutrition of Higher Plants. Academic Press, New York.

34. McComb, A. L., and Griffith, J. E. 1946. Growth stimulation and phosphorus absorption of mycorrhizal and nonmycorrhizal northern white pine and Douglas fir seedlings in relation to fertilizer treatment. Plant Physiol. 21:11-17.

35. Mejstrik, V. 1970. The uptake of ^{32}P by different kinds of ectotrophic mycorrhiza of *Pinus*. New Phytol. 69:295-298.

36. Mejstrik, V., and Benecke, U. 1969. The ectotrophic mycorrhizas of *Alnus viridis* (Chaix) D.C. and their significance in respect to phosphorus uptake. New Phytol. 68:141-149.

37. Melin, E. 1917. Studier over de norrlandska myrmarkernas vegetation. (Almqvist & Wiksells, Uppsala. Norrlandskt Handbibliotek VI Sonderbdr. A. Akad. Avhandl.

38. Melin, E., and Nilsson, H. 1950. Transfer of radioactive phosphorus to pine seedlings by means of mycorrhizal hyphae. Physiol. Plantarum 3:88-92.

39. Muller, P. E. 1878. Studier over Skovjord. som Bidrag til Skovdykningens Theori I. (Tidskr f. Skovbrug. 3).

40. Pang, P. C. K., and Holenko, H. 1986. Phosphomonoesterase activity of forest soils. Soil. Biol. Biochem. 18:35-40.

41. Rousseau, J. V. D., and Reid, C. P. P. 1987. Simulation of pine seedling response to mycorrhizae with applied phosphorus. Page 259 in: Proc. of the 7th North American Conference on

Mycorrhizae. D. M. Sylvia, L. L. Hung, and J. H. Graham, eds., Gainesville, FL.

42. Rouseau, J. V. D., Reid, C. P. P. and English, R. J. 1992. Relationship between biomass of the mycorrhizal fungus *Pisolithus tinctorius* and phosphorus uptake in loblolly pine seedlings. Soil. Biol. Biochem. 24:183-184.

43. Rygiewicz, P. T., and Bledsoe, C. S. 1984. Mycorrhizal effects on potassium fluxes by northwest coniferous seedlings. Plant Physiol. 76:918-923.

44. Skinner, M. F., and Bowen, G. D. 1974. The uptake and translocation of phosphate by mycelial strands of pine mycorrhizas. Soil Biol. Biochem. 6:53-56.

45. Stahl, E. 1900. Der Sin der Mycorhizenbildung. (Jahrb. f. wiss. Bot. 34:534-668).

46. Thomas, K. I. 1985. The utilization of inositol hexaphosphate by selected ectomycorrhizal symbionts grown in pure culture. Forestry Abstracts 46:649.

47. Tyminska, A., Le Tacon, F. and Chadoeuf, J. 1986. Effect of three ectomycorrhizal fungi on growth and phosphorus uptake of *Pinus silvestris* seedlings at increasing phosphorus levels. Can. J. Bot. 64:2753-2757.

48. Vozzo,, J. A., and Hacskaylo, E. 1971. Inoculation of *Pinus caribaea* with ectomycorrhizal fungi in Puerto Rico. For. Sci. 17:239-245.

49. White, C. S., Gosz, J. R., Horner, J. D., and Moore, D. I. 1988. Seasonal, annual and treatment-induced variation in available nitrogen pools and nitrogen-cycling processes in soils of two Douglas-fir stands. Biol. Fertil. Soils 6:93-99.

INVOLVEMENT OF CROPPING SYSTEMS, PLANT PRODUCED COMPOUNDS AND INOCULUM PRODUCTION IN THE FUNCTIONING OF VAM FUNGI

Gene R. Safir
Michigan State University
East Lansing, Michigan

It has been estimated that about 95% of the world's vascular plant species are mycorrhizal (57). Mycorrhizae are associations between fungi and plant roots. Vesicular-arbuscular mycorrhizal (VAM) associations are the most common type of mycorrhizae. There is increasing interest in these associations because they are capable, under certain situations, of greatly increasing plant growth and yield. One major reason for this increased growth and yield is the ability of VAM plants to efficiently take up nutrients, especially phosphorus. Other reasons include increasing plant tolerance to water, salt, disease, and herbicide carryover stresses. Although VAM fungi are present in most soils, it is probable that natural VAM associations are not functioning in the field in terms of increasing plant growth, for example, as efficiently as they are capable of functioning (11). This may be partially because in many field situations fungal propagule germination (43) and plant root colonization occurs over a longer time and at lower rates than can be induced under greenhouse or other controlled conditions using higher fungal inoculum concentrations. These higher fungal concentrations under controlled conditions can cause earlier and greater root colonization resulting in earlier stimulation of plant growth. Plant growth, however, may not always be stimulated by high levels of early VAM root colonization. It has been suggested that certain species of VAM fungi may sometimes decrease plant growth (24,25,26,33) particularly when high fertilization rates are used (27). Cropping sequences as well as nutritional and pest management practices also dramatically affect the VAM fungal species composition in the soil and their effects on plants. There is increasing evidence that compounds from plant roots are capable of dramatically influencing the growth and root colonization rates of VAM fungi (48). These compounds are usually phenolic and their concentration in soil is regulated to a great extent by cropping sequence.

The use of VAM fungi for practical purposes is limited by our inability to propagate the fungi in axenic culture. Inoculum for experimental purposes is usually obtained from pot cultures of the fungi grown on the roots of several plant species. This procedure is expensive, however, there are agricultural and resource management situations where the use of VAM fungal inoculum should be economical. There are numerous, recent excellent discussions of the biology and potential use of VAM fungi (1,18,19,25,41,42,53). In this chapter I will discuss the involvement of cropping systems and plant produced compounds in the functioning of VAM fungi in selected agroecosystems, as well as briefly discuss the practical utilization of VAM fungal inoculum. I will not attempt to completely review the literature relating to the above subjects and will include research from my own laboratory whenever appropriate.

PLANT GROWTH AND NUTRITION

In order for VAM fungi to be used practically in natural and agroecosystems, it is appropriate to briefly discuss how VAM function. Increased growth of many plant species in pot studies in response to root colonization by VAM fungi has been widely reported (1,2,19,42,51). Positive growth responses to VAM fungal inoculation in the field have also been reported (8,14,19,32,39,61). The plants shown to benefit from VAM inoculation include horticultural and field crops as well as desert, prairie, dune and tropical species. Examples of the plants involved are corn (*Zea mays* L.), soybean (*Glycine max* L.), grapes (*Vitis vinifera* L.), alfalfa (*Medicago sativa* L.), clover (*Trifolium repens* L.), onions (*Allium porrum* L.), asparagus (*Asparagus officinalis* L.), cowpea (*Vigna unguiculata* L.), potatoes (*Solanum tuberosum* L.), citrus (*Citrus spp.*), peach (*Prunus persica* L. Batch), cherry (*Prunus cerasus* L.), sweetgum (*Liguidambar styraciflua* L.), coffee (*Coffea arabica* L.), and maple (*Acer saccharum* Marsh). Plants generally benefit from VAM most when soil nutrients, such as P, that are needed in large amounts and are highly immobile in the soil, are present in the soil in low or moderate amounts. However, plant growth stimulation from VAM has been reported to occur over a wide range of nutrient availability, depending on the plant and fungal species, as well as soil composition and temperature. Soil factors such as water availability, pH, and salt concentration can also influence VAM formation directly, or they can act indirectly by affecting nutrient availability. For example, as soil water deficits increase, the availability of P decreases and VAM formation increases (8,37).

Most mycorrhiza researchers agree that VAM can improve plant P uptake when P availability becomes less than optimum (1,2,6,19,51) and that this increased P uptake can increase plant growth. Increasing number of studies suggest that increased P uptake by VAM plants is not totally responsible for their increased growth (39,61). The mechanisms for this growth response are not well understood. It has also been demonstrated that P and Zn, for examples, can be transported to plant roots in large quantities by VAM fungal hyphae. This suggests that the increased surface area for nutrient absorption provided by VAM fungal hyphae is a major factor in the increased nutrition of VAM plants under some conditions. Mycorrhizal and nonmycorrhizal plants are thought to utilize the same forms of soil P; however, this assumption may not be correct in all soils (48).

CROPPING SYSTEMS AND VAM

There are many demonstrated differences in the
ability of different species or strains of VAM fungi
to promote plant growth, tolerate selected
fertilization and pest management practices, and to
survive specific cropping sequences. In addition,
there are indications that the level of VAM formation
may not be a reliable indicator of the potential
benefits of VAM (30). In fact, it has been
demonstrated that some VAM fungi can be pathogenic
(21,33), and that less effective colonizers can
sometimes be better plant growth promoters than some
good colonizers (20). Perhaps the stage of
colonization, for example when and how many vesicles
or other fungal structures are formed, may influence
the plant growth response. It is logical to assume
that since vesicle formation requires large carbon
inputs from the plant, early vesicle formation might
negatively influence plant growth response under some
conditions. Suffice it to say that it is valuable to
know not only the stages, intensity, and timing of
colonization but the VAM fungal species or strains (if
possible) that are involved in any natural or
agroecosystem to be managed (25). The number of VAM
fungal species that have been identified is increasing
and over 149 species have been described. The
taxonomy of VAM fungi recently was revised (34), and
hopefully some of the past taxonomic confusion in the
literature has been eliminated.

The diversity of VAM fungal species has been shown
to decrease when agroecosystems are first formed (45).
It has been suggested by Johnson and Pfleger (25) that
the greater number of plant species in natural systems
compared to agroecosystems, plus the increased
cultural and pest management pressures in the latter,
may account for these differences. It also has been
shown that for perennial crops, such as asparagus, VAM
fungal diversity decreases continually on a seasonal
basis (63). The VAM fungi isolated from fields of
asparagus grown for 15 consecutive years did not
promote asparagus growth in the greenhouse. The
presence of high concentrations of phenolic compounds,

such as ferulic acid, in the older asparagus field
soils, was suggested to affect fungal diversity and
function (40,62). In a similar study, ferulic acid
reduced VAM formation and growth promotion of
asparagus in growth chamber experiments (62).

It has been demonstrated that leaving soils fallow
can reduce the VAM fungal population (55).
Furthermore, the results of this study strongly
suggested that the long fallow disorders of corn and
other crops are the results of decreased VAM
colonization (55,56). It was also reported (60) that
corn in corn-fallow rotations had lower early season
VAM formation rates, P contents and dry matter
production than in corn-soybean rotations. Cover
cropping, as well as precropping, also have been shown
to increase the VAM fungal inoculum potential, root
colonization, and growth of several plant species
(25).

Evidence has accumulated for centuries that crop
rotation is generally beneficial to crop production.
The reasons suggested for these benefits include
improved nitrogen nutrition when legumes are included
in the rotation, decreased disease levels, and
improved soil structure. There is now increasing
evidence that the benefits of crop rotation may be
related to population dynamics of VAM fungi (25). For
example, the decline diseases of several tree crops
(45) and perennial crops, such as asparagus (61), may
involve a shift from beneficial VAM fungal species in
a rotational system to non-beneficial species in
monocrop or perennial crop systems. Also, the
possibility has recently been raised (24) that the VAM
fungal species that predominate in monocultural corn
or soybean systems may actually cause yield
depressions. In this study, those VAM fungal species
that were predominant on corn were positively
correlated with yields and tissue nutrition of
soybeans, and negatively correlated with corn yields
and tissue nutrition. The VAM fungal species
associated with soybean were negatively correlated
with soybean yields and tissue nutrition and
positively correlated with corn yields and tissue
nutrition. Based on these correlations, several

generalizations were proposed (24) to stimulate research and discussion.

"1. Continuous cropping selects for the most rapidly growing and sporulating VAM fungal species.

2. These fungi provision their own growth at the expense of their plant symbionts and are therefore either inferior mutualists, or perhaps even parasitic, compared to non-proliferating species of VAM fungi.

3. Over time, crop vigor declines in monoculture because populations of detrimental species increase and populations of beneficial species decrease in the VAM fungal community.

4. Interruption of a monoculture (rotation) reduces the relative abundance of detrimental fungi, and increases the relative abundance of beneficial fungi."

The authors (24) recommended that these unproved generalizations should be evaluated by testing the effects of individual species in a given system in relation to crop physiology, growth and yield. In a previous study (26), these authors identified VAM fungal species that proliferated more in either corn or soybean monocropping systems. Of interest also was the observation that the percentage of roots of corn and soybean with VAM was generally unaffected by five years of previous monocropping of either corn or soybean. These results are supported by another study (60) which showed that the VAM levels in corn throughout the season following either corn or soybean in rotation were similar. If the first generalization presented (24) is correct, we must determine how similar colonization levels in corn-corn rotations, for example, can possibly have neutral or negative influences on yield, compared to corn in a corn-soybean rotation. In some way, perhaps, VAM in the corn-corn system cause a larger carbon drain from corn. In all of these studies, only percent (%) of roots with VAM was measured. However, there are situations where measurement of intensity (stage of VAM colonization, for example, how fast and how many vesicles or other fungal structures are formed in the root) (37) or external hyphal development (12) of VAM

formation can be more meaningful than percent root
colonization. Also, another study suggested that
increased VAM fungal colonization is not a necessary
component of the increased plant growth observed in
undisturbed soil (31). Thus, differences in
experimental methods are of critical importance when
comparing low levels of VAM fungal colonization in
early plant growth that are not greatly different. We
have seen, for example, a treatment with a given
percentage of VAM fungal colonization has either
higher or lower intensity rating than another
treatment with the same percentage VAM fungal
colonization. We have also seen a treatment with a
higher percent VAM fungal colonization have lower
intensity rates (37). As more field studies involving
VAM fungal systems are conducted, it is becoming clear
that the relationship between development of the VAM
fungal association and plant performance is more
complicated than pot studies have suggested. For
example, it may be of critical importance to consider
many additional parameters if our understanding of VAM
fungal systems is going to lead to their management
and/or control. These additional parameters range
from VAM fungal turnover in soil and roots, root
turnover and longevity, inter-plant hyphal connections
and grazing of external mycelium (11).

Very little information is available concerning
the effects of differences in VAM fungal species
diversity and function in relation to cropping
practices, with the possible exception of pesticide
related effects (25,58). It has been shown that a
wide range of pesticides can alter germination of VAM
fungal spores, growth, and root colonization. Until
the mechanisms controlling VAM formation are more
fully understood, it will not be possible to
effectively manage the indigenous VAM fungal
populations in most agroecosystems.

ROOT EXUDATES AND PLANT RESIDUES

In recent years, there has been increased
interest in the possible role of root exudates and
plant residues in the regulation of VAM fungal growth

and root colonization (48). It is possible that, as with other symbiotic systems, root exudates and residues may provide clues which will enable the management and possibly the pure culture of VAM fungi. In addition, it is possible that stimulatory compounds from plant roots can stimulate the indigenous populations of VAM fungi in the field and thereby increase plant growth and yields.

It has been suggested (9,17,64) that phenolic acids leached from plant residues into the soil can reduce crop yields by affecting plant growth directly or indirectly through chemically-mediated changes in the soil microbial community. There is now limited evidence that phenolic acids such as ferulic, caffeic, and methylenedioxy-cinnamic (MDC) can reduce VAM formation and decrease the VAM plant growth response. Since these compounds increased in asparagus fields to levels toxic to VAM formation, it was suggested that a chemically induced decrease in VAM may account for part of the asparagus decline syndrome (40,62). Additional unpublished and on-going studies by L. Fries and G. R. Safir have shown that p-coumaric acid, p-hydroxybenzoic acid and quercetin could reduce VAM levels in clover when 1.0 mM solutions were applied every 4 days; p-coumaric acid and quercetin at the same concentration had the same effect on sorghum.

It was shown as early as 1982 (15) that root exudates could stimulate VAM fungal spore germination. It has also been suggested that high soil P concentrations can inhibit VAM formation by decreasing root exudation (16). Characterization of the root exudates from P deficient roots compared with those from non-deficient roots, however, indicated that there were no detectable qualitative differences. We found, however, that only the exudates from P stressed roots stimulated hyphal growth of a VAM fungus (INVAM 112) on artificial media (10). Further research regarding the nature of the stimulatory root exudates led to the isolation, purification and synthesis of the compounds involved. The two compounds involved were the isoflavonoids, formononetin and biochanin A (36). Formononetin also stimulated hyphal growth of *Gigaspora gigantea* (Nicolson and Gerdemann) Gerd. and

Trappe in cell suspension culture (J. O. Siquiera, personal communication). Other researchers have tested flavonoid compounds for their effects on VAM fungal spore germination and hyphal growth (4,5,13,59). Two studies (4,59) revealed that the flavonol quercetin promoted VAM fungal spore germination and hyphal growth. However, no stimulation of hyphal growth or spore germination with any isoflavonones, such as formononetin or biochanin A, occurrred. Also, in the study by Becard et al. (4) the flavonones hesperitin and naringenin, which stimulated spore germination and hyphal growth in another study were ineffective (13). While these different results remain unexplained it should be noted that different VAM fungal species, differences in media (which dramatically affect the availability of the compounds), and different growing conditions could account for the contrasting results. Also, both VAM fungal spore germination and hyphal growth were dramatically increased when carbon dioxide concentrations were raised to 2% in the incubators (4), concentrations not uncommon in the field.

We tested the compounds formononetin and biochanin A for their possible effects on VAM formation. Both compounds increased VAM formation and growth of white clover (*Trifolium repens* L.) when added to soil at concentrations of 5 ppm (50), but were not stimulatory at 20 ppm. Several other compounds were tested with mixed results. For example, the flavone, chrysin, at much higher concentrations increased VAM colonization, but to a lesser extent than did formononetin. The effects were dependent on VAM fungal spore and compound concentrations as well as stage of plant growth; no stumulation occurred at both high inoculum levels and high formononetin concentrations. Additional studies have demonstrated that application of formononetin to the soil can increase VAM formation and plant growth for both corn and sorghum, but did not stimulate VAM to form on nonmycorrhizal canola (*Brassica campestris* L.) (49). We have also shown that the application of formononetin or biochanin A to soils containing indigenous or applied VAM fungi can enable corn and sorghum to better tolerate toxic levels of the herbicide imazaquin. Interestingly, the addition of high levels of VAM fungal spores increased VAM levels and decreased the toxicity of the applied herbicide. Preliminary field trials in which formononetin was applied to the soil before sowing with corn resulted in growth and yield increases in Michigan (G. R. Safir, unpublished), however, these results have not been confirmed. Additional field studies in Brazil in 1992, showed 12-45% increases in yield of corn and soybean, depending on concentration of formononetin added (47).

Our current research is focused on determining the mechanism of action of selected root exudates in relation to increased VAM formation and plant growth. Of particular interest are fungal uptake characteristics and the effects of the isoflavone formononetin on isoenzyme activity. In this regard, A. Ozan, R. Pacovsky and I (unpublished) have found a specific VAM fungal isozyme of malate dehydrogenase whose activity increased to varying degrees by formononetin, depending on the stage and intensity of VAM formation. Additional differences in isozyme activity between VAM and nonVAM roots have been found for peroxidase and succinate dehydrogenase, and some of these responses are modified by formononetin. Studies are also in progress to determine flavonoid breakdown rates and mobility in soil. These studies hopefully will lead to a formulation of formononetin, other than a simple water solution, that can be used for additional research and field testing. Formononetin breaks down in the soil more slowly than biochanin A, for example, and the breakdown rates are increased by soil microbial activity and the presence of plants. As would be expected, the mobility of formononetin in the soil water solution is much greater than that of solid or powdered formononetin dispersed by irrigation.

INOCULUM PRODUCTION

In order for VAM fungi to be applied to agricultural systems, it is necessary to produce sufficient quantities of inoculum. There are many characteristics that an effective VAM fungal inoculum should possess, and this subject was recently reviewed extensively (23). VAM fungal inoculum must promote economically significant growth or stress tolerance, be free of pathogens and contaminants, be economically produced and formulated, with sufficient shelf life. There are several methods available to produce VAM fungal inoculum. The most common technique used to grow VAM fungal inoculum is the pot culture method wherein roots or soil containing the fungus, are used to inoculate plants grown in pasteurized soil. After several months, roots are sufficiently colonized that the resulting spores, infected roots, and soil containing infective fungal propagules can be used as inoculum. Soilless media such as bark, peat, perlite, and expanded clay aggregates have also been used for inoculum increase (29). Inoculum production in these media usually is highest at low to moderate soil P levels and under conditions favoring rapid plant growth. Additional methods for producing inoculum include hydroponic (35) systems, and a modified system called the nutrient film technique (NFT) in which colonized plant roots are bathed by a thin film of nutrient solution, allowing for good aeration. An aeroponic culture method, which involves applying a fine mist of nutrient solutions to colonized roots, has also been modified for VAM fungal inoculum production (22,52). The aeroponic method has been shown to produce high levels of contaminant free fungal spores as well as highly infective root fragments that remain infective for up to 23 months (54). Additionally, inoculum levels as high as 135,000 fungal propagules/gram of dried roots have been produced in the aeroponic system using a shearing-sieving process. Root organ culture techniques may also be useful to produce clean VAM fungal inoculum (65) especially if roots are transformed by inoculation with *Agrobacterium rhizogenes* which increases the rate of root production and thus quantities of VAM fungal propagules. In fact, this method has been modified to produce even larger quantities of fungal propagules by inoculating large volumes of autoclaved peat-vermiculite mixture with roots from agar culture in order to grow large quantities of roots (65).

If the inoculum produced by any of the methods described above is to be effective, sufficient quantities of VAM fungal propagules must be placed in such a way to ensure contact with initial root growth or radicle emergence. For field crops grown from seed, the most successful inoculations have resulted from inoculum placement about one inch below the seed. Seed treated with VAM fungi is another method of inoculation and subsequent colonization of the emerging radicle. Placement of fungal inoculum for container grown plants is less critical since the roots should eventually come in contact with the fungus, regardless of placement.

The methods described above have all provided VAM fungal inoculum for research purposes. In addition, at least two commercial VAM fungal inoculum products are in production at this time: Mycori Mix (produced by Premier Peat Moss LTD. Riviere-du- Loup, Quebec, Canada), and Dr.Kinkon (produced by Idemitsu Kosan Co. LTD, Japan). Because of economic considerations, high value horticultural crops are likely to benefit the most from the use of commercial VAM fungal inoculants. The production systems for many of these crops are readily modifiable for the introduction of the VAM fungus in relatively small quantities, compared to amounts required for field crop systems grown from seed. It is also probable that the benefits of VAM fungal inoculation for horticultural crops are likely to be more predictable than for field crops. The reasons are related to existing horticultural procedures that employ micropropagation, soilless media, and disinfested soil.

VAM FUNGAL INOCULATION: A CASE STUDY

The best way to describe VAM application technology is to present a brief economic analysis related to benefits of VAM inoculation in a specific system. Cold climate grafted grapevines in Michigan and New York must grow for 3-5 years after planting before an economically significant yield of grapes is achieved. A winemaker will pay about $1500/ton for high quality chardonnay grapes. The average yield for chardonnay in Northern Michigan is about 3 tons per acre. We inoculated chardonnay grapevines with Mycori Mix in two Northern Michigan commercial vineyards three years ago. Inoculation increased vine growth for the first two years by about 40%, and increased the grape yields after the third season by 24%, compared with the yields of vines planted without inoculation. The increased income/acre would amount to about $1000; the retail cost of inoculum would be about $150/acre; clearly economically justifiable. Obviously, an inoculum cost of $150/acre would be unacceptable to a corn or soybean grower where the limit would probably be less than $10 per acre total for both inoculum and inoculation (delivery to root zone). For these systems a high quality, highly concentrated VAM fungal inoculum formulation is probably essential. Also, a mixture of VAM fungi may prove to be more effective than a formulation containing only one fungus.

CONCLUSIONS

VAM are the most prevalent types of mycorrhizal associations and they occur on most crop plants grown throughout the world. The demonstrations that VAM stimulate growth and yields of numerous crops has encouraged a great deal of research and commercial interest in these systems. The current trends in agriculture relating to low chemical input sustainable systems may be contributing to this interest.

The role that VAM play in various cropping systems is being studied more intensely, and although this research is time consuming and expensive, our knowledge of the VAM role is increasing. It is apparent that fallow rotations and certain pesticide regimes can reduce VAM effectiveness on some crops grown the following season. It has also been suggested that VAM can cause negative or neutral effects on plant growth in some perennial or monocropping systems, possibly due to shifts in VAM fungal species. Little information is available concerning the mechanisms of negative effects of VAM, however it has been suggested that allelopathic phenolic compounds may affect both the VAM fungal diversity and function in perennial systems. Several studies have shown that flavonoid root exudates can dramatically increase or decrease VAM fungal hyphal growth, root colonization and subsequent plant growth. It is possible, therefore, that the flavonoids are stimulating the indigenous populations of VAM fungi in the field with a resulting increase in yield. Initial field tests with the isoflavone, formononetin, resulted in increased yields of both corn and soybean.

The successful use of VAM fungal inoculum in the field has been demonstrated for many crops, however, our inability to grow VAM fungi axenically has limited the development of economically and biologically effective inoculum for many agronomic systems. Horticultural systems are likely to benefit the most initially from VAM systems because production practices of these high value crops are readily amenable to inoculation. Current inoculum development technologies are capable of producing sufficient quantities of effective inoculum with adequate shelf life for economical use in many horticultural systems. Two commercial VAM fungal inoculum formulations are available.

LITERATURE CITED

1.   Abbott, L. K., and Robson, A. D. 1991. Factors influencing the occurrence of vesicular-arbuscular mycorrhizas. Agric. Ecosystems Environ. 35:121-150.
2.   Amijee F., Tinker, P. B., and Stribley, D. P. 1989. The development of endomycorrhizal root systems VII. A detailed study of effects of soil phosphorus on colonization. New Phytol. 111:435-446.
3.   Azcon, R., and Ocampo, J. A. 1984. Effect of root exudation on VA infection at early stages of plant growth. Plant Soil 82:133-138.
4.   Becard, G., Douds, D. D., and Pfeffer, P. E.

1992. Extensive in vitro hyphal growth of vesicular mycorrhizal fungi in the presence of CO_2 and flavonols. Appl. Environ. Microbiol. 58:821-825.

5. Becard, G., and Piche, Y. 1989. New aspects on the acquisition of biotrophic status by a vesicular-arbuscular mycorrhizal fungus *Gigaspora margarita*. New Phytol.112:77-83.

6. Black, R. L. B., and Tinker, P. B. 1977. Interaction between effects of vesicular-arbuscular mycorrhiza and fertilizer phosphorus on yields of potato in the field. Nature 267:510-511.

7. Black, R. L. B., and Tinker, P. B. 1979. The development of endomycorrhizal root systems. II. Effect of agronomic factors and soil conditions on the development of vesicular-arbuscular mycorrhizal infection in barley and on the endophyte spore density. New Phytol. 83:401-413.

8. Bolgiano, N. C., Safir, G. R., and Warncke, D. D. 1983. Mycorrhizal infection and growth of onion in the field in relation to phosphorus and water availability. J. Amer. Soc. Hort. Sci. 108: 819-825.

9. Dalton, B. R. 1989. Physicochemical and biological processes affecting the recovery of exogenously applied ferulic acid from tropical forest soils. Plant Soil 115: 13-22.

10. Elias, K. S., and Safir, G. R. 1987. Hyphal elongation of *Glomus fasciculatus* in response to root exudates. Appl. Environ. Microbiol. 53:1928-1933.

11. Fitter, A. H. 1985. Functioning of vesicular-arbuscular mycorrhizas under field conditions. New Phytol. 99:257-265

12. Gazey, C., Abbott, L. K., and Robson, A. D. 1992. The rate of development of mycorrhizas affects the onset of sporulation and production of external hyphae by two species of Acaulospora. Mycol.Res. 96:643-650

13. Gianinazzi-Pearson, V., Branzanati, V. B., and Gianinazzi, S. 1989. In vitro enhancement of spore germination and early hyphal growth of a

vesicular-arbuscular mycorrhizal fungus by host root exudates and plant flavonoids. Symbiosis 7:243-255.

14. Gianinazzi, S., Trouvelot, A., and Gianinazzi-Pearson, V. 1990. Role and use of mycorrhizas in horticultural crop production. Pages 25-30 in: Proc. XXII International Horticultural Congress (I.S.H.S.), Int.Soc. for Hort. Science, Aug.-Sept. 1, 1990.

15. Graham, J. H. 1982. Effect of citrus root exudates on germination of chlamydospores of the VA-mycorrhizal fungus *Glomus epigaeum*. Mycologia 74:831-835.

16. Graham, J. H., Leonard, R. T., and Menge, J. A. 1981. Membrane mediated decrease in root exudation responsible for phosphorus inhibition of vesicular-arbuscular mycorrhiza formation. Plant Physiol. 68:548-552.

17. Guenzi, W. D., and Mc Calla, T. M. 1966. Phenolic acids in oats, wheat, sorghum, and corn residues and their phytotoxity. Agron.J. 58:303-304.

18. Harley, J. L., and Smith, S. E. 1983. Mycorrhizal symbiosis. Academic Press, New York, NY. 483 pp.

19. Hayman, D. S. 1987. VA mycorrhizas in field crop systems. Pages 171-192 in: Ecophysiology of VA-mycorrhizal plants. G.R. Safir, ed., CRC Press, Boca Raton, Fla.

20. Hetrick, B. A. D., and Wilson, G. W. T. 1991. Effects of mycorrhizal fungus species and metalaxyl applications on microbioal suppression of mycorrhizal symbiosis. Mycologia 83:97-102.

21. Howeler, R. H., Sieverding, E., and Saif, S. 1987. Practical aspects of mycorrhizal technology in some tropical crops and pastures. Plant Soil 100:249-283.

22. Hung, L. L., and Sylvia, D. M. 1988. Production of vesicular-arbuscular mycorrhizal fungus inoculum in aeroponic culture. Appl. Environ. Microbiol. 54:353-357.

23. Jarstfer, A. G., and Sylvia, D. M. 1992. Inoculum production and inoculation technologies of vesicular-arbuscular mycorrhizal fungi. in: Soil Microbial Ecology: Applications in agriculture,

forestry and environmental management B. Metting, ed., Marcel Dekker, Inc., New York. Florida Ag. Expt. St. J. Ser. No. R-01157.

24. Johnson, N. C., Copeland, P. J., Crookston, R. K., and Pfleger, F. L. 1992. Mycorrhizae: Possible explanation for yield decline with continuous corn and soybean. Agron. J. 84:387-390.

25. Johnson, N. C., and Pfleger, F. L. 1992. Vesicular-arbuscular mycorrhizal and cultural stresses. Pages 71-99 in: Mycorrhizae in Sustainable Agriculture. G. J. Bethlenfalvay and R. G. Linderman, eds., ASA Special Publication No.54, Madison, WI.

26. Johnson, N. C., Pfleger, F. L., Crookston, R. K., Simmons, S. R., and Copeland, P. J. 1991. Vesicular-arbuscular mycorrhizas respond to corn and soybean cropping history. New Phytol. 177:657-663.

27. Kiernan, J. M., Hendrix, J. W., and Maronek, D. M. 1983. Fertilizer-induced pathogenicity of mycorrhizal fungi to sweetgum. Soil Biol. Biochem. 15:257-262.

28. Lambert, D. H., Cole, H., and Baker, D. E. 1980. Adaptation of vesicular-arbuscular mycorrhizae to edaphic factors. New Phytol. 85:513-520.

29. Liyanage, H. D. 1989. Effects of phosphorus nutrition and host species on root colonization and sporulation by vesicular-arbuscular mycorrhizal fungi in sand-vermiculite medium. M.S.thesis, Univ. of Florida, Gainesville.

30. McGonigle, T. P. 1988. A numerical analysis of published field trials with vesicular-arbuscular mycorrhizal fungi. Functional Ecology 2:473-478.

31. McGonigle, T. P., Evans, D. G., and Miller, M. H. 1990. Effect of degree of soil disturbance on mycorrhizal colonization and phosphorus absorption by maize in growth chamber and field experiments. New Phytol. 116:629-636.

32. Menge, J.A., Raski, D. J., Lider, L. A., Johnson, E. L. V., Jones, N. O., Kissler, J. J., and Hemstreet, C. L. 1983. Interactions between mycorrhizal fungi, soil fumigation, and the

growth of grapes in California. J. Enol. Vitic. 34:117-121.

33. Modjo, H. S., and Hendrix, J. W. 1986. The mycorrhizal fungus *Glomus macrocarpum* as a cause of tobacco stunt disease. Phytopathology 76:688-691.

34. Morton, J. B., and Benny, G. L. 1990. Revised classification of arbuscular mycorrhizal fungi (*Zygomycetes*): A new order, *Glomales*, two new suborders, *Glomineae* and *Gigasporineae*, and two new families, *Acaulosporaceae* and *Gigasporaceae*, with an emendation of *Glomaceae*. Mycotaxon 37:471-491.

35. Mosse, B., and Thompson, J. P. 1984. Vesicular-arbuscular endomycorrhizal inoculum production. I. Exploratory experiments with beans (*Phaseolus vulgaris*) in nutrient flow culture. Can. J. Bot. 62:1523-1530.

36. Nair, M. G., Safir, G. R., and Siqueira, J. O. 1991. Isolation and identification of vesicular-arbuscular mycorrhiza-stimulatory compounds from clover (*Trifolium repens*) roots. Appl. Environ. Microbiol. 57:434-439.

37. Nelsen, C. E., and Safir, G. R. 1982. Increased drought tolerance of mycorrhizal onion plants caused by improved phosphorus nutrition. Planta 154:407-413.

38. Ocampo, J.A., and Hayman, D. S. 1981. Influence of plant interactions on vesicular-arbuscular mycorrhizal infections. II. Crop rotations and residual effects of non-host plants. New Phytol. 87:333-343.

39. Pedersen, C. T., Safir, G. R., Parent, S., and Caron, M. 1991. Growth of asparagus in a commercial peat mix containing vesicular-arbuscular mycorrhizal (VAM) fungi and the effects of applied phosphorus. Plant Soil 135:75-82.

40. Pedersen, C. T., Safir, G. R., Siqueira, J. O., and Parent, S. 1991. Effects of phenolic compounds on asparagus mycorrhiza. Soil Biol. Biochem. 23:491-494.

41. Powell, C. L., and Bagyaraj, D. J. 1984. VA

Mycorrhiza. CRC Press, Boca Raton, Fla. 234 pp.

42. Safir, G. R. 1987. Ecophysiology of VA-mycorrhizal plants. CRC Press, Boca Raton, Fla. 224 pp.

43. Safir, G. R., Coley, S. C., Siqueira, J. O., and Carlson, P.S. 1991. Improvement and synchronization of VA mycorrhiza fungal spore germination by short term cold storage. Soil Biol. Biochem. 109:109-111.

44. Schenck, N. C., and Siqueira, J. O. 1987. Ecology of VA mycorrhizal fungi in temperate agroecosystems. Pages 2-4 in: Mycorrhizae in the next decade- practical applications and research priorities. Proc. 7th North Amer. Conf. on Mycorrhizae, D. Sylvia, L. Hung and J. Graham, eds., Univ. of Florida, Gainesville.

45. Schenck, N. C., Siqueira, J. O., and Oliveira, E. 1989. Changes in incidence of VA mycorrhizal fungi with changes in ecosystems. Pages 125-129 in: Interrelationships between microorganisms and soil. V. Vancura and F. Kunc, eds., Elsevier, New York, N.Y.

46. Schwab, S. M., Leonard, R. T., and Menge, J. A. 1984. Quantitative and qualitative comparison of root exudates of mycorrhizal and nonmycorrhizal plant species. Can. J. Bot. 62:1227-1231.

47. Siqueira, J. O., Brown, D. G., Safir, G. R., Nair, M. G. 1992. Field applications of the VA-mycorrhiza stimulating isoflavonoid formononetin (Rhizotropin™) on corn and soybean in Brazil. Page 133 in: The Int. Symp. on Management of Mycorrhizas in Agriculture and Forestry, The Univ. of Western Australia, Nedlands, Western Australia, Sept.28- Oct.2, 1992.

48. Siqueira, J. O., Nair, M. G., Hammerschmidt, R., and Safir, G. R. 1991. Significance of phenolic compounds in plant-soil-microbial systems. Crit. Rev. in Plant Sciences 10:63-121.

49. Siqueira, J. O., Safir, G. R., and Nair, M. G. 1991. VA-mycorrhizae and mycorrhiza stimulating isoflavovoid compounds reduce plant herbicide injury. Plant Soil 134:233-242.

50. Siqueira, J .O., Safir, G. R., and Nair, M. G.

1991. Stimulation of vesicular-arbuscular mycorrhiza formation and growth of white clover by flavonoid compounds. New Phytol. 118:87-93.

51. Stribley, D. P. 1987. Mineral nutrition. Pages 59-70 in: Ecophysiology of VA mycorrhizal plants. G.R. Safir, ed., CRC Press, Boca Raton, Fla.

52. Sylvia, D. M., and Hubbell, D. H. 1986. Growth and sporulation of vesicular-arbuscular mycorrhizal fungi in aeroponic and membrane systems. Symbiosis 259-267.

53. Sylvia, D. M., and Williams, S. E. 1992. Vesicular-arbuscular mycorrhizae and environmental stresses. Pages 101-124 in: VA Mycorrhizae in Sustainable Agriculture, G. J. Bethlenfalvay and R. G. Linderman, eds., ASA Special Public No.54. Madison, WI.

54. Sylvia, D. M., and Jarstfer, A. G. 1992. Sheared-root inocula of vesicular-arbuscular mycorrhizal fungi. Appl. Environ. Microbiol. 58:229-232.

55. Thompson, J. P. 1987. Decline of vesicular-arbuscular mycorrhizal in long fallow disorder of field crops and its expression in phosphorus deficiency of sunflower. Austr. J. Agric. Res. 38:847-867.

56. Thompson, J. P. 1991. Improving the mycorrhizal condition of the soil through cultural practices and effects on growth and phosphorus uptake by plants. Pages 117-137 in: Phosphorus nutrition of grain legumes in the semi-arid tropics. Int. Crops Res. Inst. for the semi-arid Tropics (ICRISAT), C. Johansen et al., eds. Patancheru, India.

57. Trappe, J. M. 1987. Phylogenetic and ecological aspects of mycotrophy in the angiosperms from an evolutionary standpoint. Pages 5-25 in: Ecophysiology of VA Mycorrhizal plants, G. R. Safir, ed., CRC Press, Boca Raton, FL.

58. Trappe, J. M., Molina, R., Castellano, M. 1984. Reactions of mycorrhizal fungi and mycorrhiza formation to pesticides. Annu. Rev. Phytopathol. 22:331-359.

59. Tsai, S. M., and Phillips, D. A. 1991. Flavonoids released naturally from alfalfa

promote development of symbiotic *Glomus* spores in vitro. Appl. Environ. Microbiol. 57:1485-1488.

60. Vivekanandan, M., and Fixen, P. E. 1991. Cropping systems effects on mycorrhizal colonization, early growth and phosphorus uptake of corn. Soil Sci. Soc. Am. J. 55:136-142.

61. Wacker, T. L., Safir, G. R., and Stephens, C. T. 1990. Effect of *Glomus fasciculatum* on the growth of asparagus and the incidence of Fusarium root rot. J. Amer. Soc. Hort. Sci. 115:550-554.

62. Wacker, T. L., Safir, G. R., and Stephens, C. T. 1990. Effects of ferulic acid on *Glomus fasciculatum* and associated effects on phosphorus uptake and growth of asparagus (*Asparagus officinalis*) J. Chem. Ecol. 16:901-909.

63. Wacker, T. L., Safir, G. R., and Stephenson, C. T. 1989. Evidence for succession of mycorrhizal fungi in Michigan asparagus fields. Acta Horticulturae 271:273-278.

64. Wang, T. S. C., Yang, T. K., Chuang, T. T. 1967. Soil phenolic acids as plant growth inhibitors. Soil Sci. 103:239-245.

65. Wood, T. 1987. Commercial production of VA-Mycorrhizal inoculum. Page 274 in: Mycorrhizae in the next decade - practical applications and research priorities, D. M. Sylvia, L. L. Hung, and J. H. Graham, eds., Univ. of Florida, Gainesville.

CURRENT STATUS OF OUTPLANTING STUDIES USING ECTOMYCORRHIZA-INOCULATED FOREST TREES

Michael A. Castellano
USDA-Forest Service
Pacific Northwest Research Station,
Corvallis, Oregon

All temperate forest trees depend on mycorrhizae for establishment and development. Commercially important ectomycorrhizal tree species belong to an array of plant families; e.g., Betulaceae, Dipterocarpaceae, Fagaceae, Myrtaceae, and Pinaceae. The endomycorrhizal fungi colonize forest trees of many other families including the Aceraceae, Taxaceae, Taxodiaceae, and Cupressaceae; to some extent the Myrtaceae; and many tropical families. Effects of endomycorrhizal inoculation on field responses of forest trees either in native forests or exotic plantations have not been studied experimentally. The earliest studies on outplanting performance involved inoculation of trees with soil from under established forests. This "mycorrhizal" inoculum also contained other rhizosphere organisms that can stimulate seedling growth (11,12,20).

The mycorrhizal habit was described in the mid-1880's (14,15,16) but received only limited attention for many years. On average through 1920 less than 20 papers were published per year on various types and aspects of mycorrhizae. Even through 1950, there still were less than 80 papers per year, and not until the late 1970's to the present did research on many different aspects and all types of mycorrhizae blossom (5,40). This trend mirrors the attention paid to outplanting performance (3,4).

The purposes of this chapter are to briefly review results of published studies on plantation response of

forest trees to ectomycorrhizal inoculation, outline
new and continuing research projects investigating
outplanting response to ectomycorrhizal inoculation,
summarize inoculum types and inoculation techniques,
and propose future priorities for research. It is
beyond the scope of this chapter to review in detail
all the literature on outplanting studies of
mycorrhiza-inoculated seedlings. Selected references
are provided as examples to lead the reader to
additional literature.

SUMMARY OF PUBLISHED STUDIES

The importance of ectomycorrhizal inoculation for
successful establishment of forest tree seedlings was
first reported in 1927 by Kessell (17) for *Pinus
radiata* D. Don grown in Australia. *Rhizopogon
luteolus* Fries sporocarps were associated with fast
growing healthy *P. radiata* seedlings, whereas stunted
seedlings had no associated sporocarps. During the
next few decades, there were only a few scattered
reports on the outplanting performance of different
tree species inoculated with various mycorrhizal fungi
(23,30,31). Not until the 1970's did intensive
research on *Pisolithus tinctorius* (Pers.) Coker &
Couch by D. H. Marx and others at the USDA Forest
Service Institute for Mycorrhizal Research and
Development engender wide interest in the
forest research community.

In 1966, Schramm (39) reported the widespread
association of *P. tinctorius* with pines established
and thriving on harsh coal spoils in the eastern
United States. Over the next 25 years, Marx and
others (24) conducted extensive experimental work on
the physiological aspects of the mycorrhizal
association of *P. tinctorius* with various conifer and
hardwood species. During their investigations,
techniques were developed for rapid and consistent
inoculation of nursery stock (22,38). This led to the
first comprehensive field testing of any mycorrhizal
fungus across a range of field conditions and host
species. *Pisolithus tinctorius* inoculation and
outplanting trials were conducted on numerous conifer

and hardwood species in Australia, Brazil, Canada,
China, Congo, France, Ghana, India, Liberia, Malawi,
Mexico, Nigeria, Puerto Rico, South Korea, Thailand,
Venezuela (13,19,24,41), and many states in the United
States (22,37,38). Typically when *P. tinctorius* was
inoculated onto exotic host species, growth was
significantly improved over noninoculated plants.
Marx (23,24) reviewed these programs in detail.

In contrast, *P. tinctorius* is not as effective
when inoculated on conifer seedlings in the Pacific
Northwest (8). Subsequently, the USDA Forest Service
in the Pacific Northwest developed the use of
basidiospores of *Rhizopogon* species as mycorrhizal
inoculum for Douglas-fir (*Pseudotsuga menziesii*
(Mirb.) Franco) and some pine species (2,6,7).
Douglas-fir seedlings inoculated with *Rhizopogon
vinicolor* Smith or *R. colossus* Smith survived and grew
better than noninoculated seedlings under a wide range
of outplanting conditions in the Pacific Northwest and
England (2,42).

Over 100 papers published on outplanting
performance of ectomycorrhizal fungus-inoculated tree
seedlings are scattered across dozens of journals,
resulting in considerable confusion regarding methods
and results. Castellano (3,4) and O'Dell et al. (36)
reviewed the literature and found highly variable
effects, some studies reporting outplanting success,
others failure, and still others mixed results.
Inoculation methodology differed substantially from
one study to another, and many studies lacked rigor in
experimental design, comprehensiveness of the measured
parameters, and statistical analysis. Often, the
percentage of seedlings with mycorrhizae and
percentage of feeder roots with mycorrhizae per
seedling were not reported.

Few mycorrhiza-inoculation and outplanting studies
have involved species from major ectomycorrhizal plant
families such as Caesalpiniaceae, Casuarinaceae,
Dipterocarpaceae, Fagaceae, Mimosaceae, and Myrtaceae.
Most tree species studied are in the Pinaceae, with
little representation of native plant species outside
North America. Pines were emphasized, especially
species from the southeastern United States planted as

exotics elsewhere in the world.

Pisolithus tinctorius, the most frequently inoculated fungus, has been used in 40% of the outplanting trials. Less than half of the trials involving *P. tinctorius* showed an improvement in seedling performance due to inoculation. Benefits of inoculating seedlings with *P. tinctorius* were best shown when seedlings were planted as exotics or planted on harsh sites such as coal mine spoils. *Pisolithus tinctorius* forms mycorrhizae with many tree species and is particularly well adapted to warm, dry, acidic soils. How *P. tinctorius* would compare to other native, host-specific fungi such as *Rhizopogon* spp. is unknown.

NEW AND ONGOING OUTPLANTING STUDIES

Considerable research on effects of ectomycorrhizal fungal inoculation on outplanting performance of forest seedlings is now in progress. Table 1 lists new and ongoing studies. New studies refer to unpublished research in progress; ongoing studies have been published but remain active in gathering additional field data on response of seedlings through time. The reader is encouraged to directly contact the scientist involved for further details of those studies.

Eight genera and 39 species of host plants are currently under investigation (Table 1). As is the case for published studies (3), about half of the new and ongoing studies involve *Pinus*. Plants have been inoculated with one of 33 fungal species belonging to 16 genera. The fungus most commonly investigated was *P. tinctorius*, being the mycobiont in about 38% of all studies. Sixteen scientists contributed information on the 191 fungus-host-location combination trials in 12 different countries, at least 5 States in the U.S., and 4 provinces in Canada.

FORMS OF INOCULUM AND INOCULATION TECHNIQUES

Efficacious ectomycorrhizal fungal inoculum in various forms is necessary to support the varied and

specific needs of a mycorrhiza management program.
Inoculum form and inoculation techniques that allow
flexibility are requisite for success. Tree seedlings
are raised in relatively controlled environments in
bareroot and container nurseries. Each nursery
differs operationally from others to some degree, even
within bareroot or container operations. Host plants
also differ drastically in biology and physiology,
thus their response to mycorrhiza management is
impacted.

Soil, nurse seedlings, spores and vegetative
mycelium are the four primary sources of
ectomycorrhiza inoculum for forest tree nurseries.
Each has advantages and disadvantages depending on
local economics and logistics. Detailed procedures
for inoculum production and nursery inoculation are
discussed by Castellano and Molina (6), Marx and
Kenney (28), Marx et al. (25), Marx et al. (26) and
Molina and Trappe (33). Modification and adaptation
of these techniques are ongoing and may improve the
likelihood of success in specific situations.

Soil

Soil transfer from established forests or
plantations to the nursery is common in developing
countries. The uppermost soil layer(s) is removed
from beneath thriving trees of the same genus as grown
in the nursery, transported to the nursery and dressed
onto the nursery beds or incorporated into the
substrate prior to seeding. In bareroot nurseries up
to 10% by volume of soil inoculum is incorporated into
the top 10 cm of the nursery beds. Soil transfer is
cost effective if labor is plentiful and inexpensive.
Soil is bulky and large amounts are needed on a
continual basis. Identity of the mycobiont is
generally unknown and can vary from year to year. The
impact on productivity of the site from which the soil
is removed is unknown, but is probably detrimental
over time. Alternatively, soil can be moved from bed
to bed in bareroot nurseries, once the mycorrhizal
fungus has been established. The potential for
transferring pathogenic organisms and weed propagules
in soil could be a significant risk and may limit the

usefulness of this procedure.

Nurse Seedlings

This method of inoculation involves transplanting established seedlings that have abundant mycorrhizae from the forest, plantation, or older beds in the nursery into newly planted beds. The mycorrhizal fungus grows from the nurse tree and colonizes the surrounding seedlings in an ever increasing concentric circle, but usually colonization of adjacent seedlings is slow and uneven. Identity of the mycobiont may be unknown and can vary from year to year. The relatively large transplant seedlings can compete with young seedlings for light, water and nutrients. This method requires availability of older established seedlings and will disrupt nursery operations (sowing, cultivating, and root pruning) because of nonuniformity of the crop. Nurse seedlings are only applicable in bareroot nurseries since spread of the inoculum is through the substrate unless the mycobiont produces fruiting bodies and spores for dispersal.

Vegetative Mycelium

Preparation of vegetative inoculum is expensive and time-consuming. Sporocarps should be kept as voucher specimens and expertly identified and preserved in a recognized herbarium. To establish cultures, tissue explants are aseptically removed from sporocarps and placed on agar media in tubes or petri plates. Modified Melin-Norkrans (MMN) or potato dextrose agar are most commonly used (32). Once the fungus is successfully cultured, hyphae from the margin of the colony are transferred into a container partially filled with MMN nutrient solution. This starter culture is allowed to grow for several weeks to months and is used to inoculate liquid media in flasks, or autoclavable bags partially filled with peat-vermiculite substrate amended with MMN nutrient solution. After 3-4 months growth, the colonized substrate is leached with water to remove excess nutrients and air-dried over-night. The inoculum is then used to inoculate nursery substrate at a desired concentration of 1 part inoculum to 10-60 parts

substrate, depending on fungus used and substrate
conditions. Inoculum can be banded beneath or beside
seed or broadcast and incorporated into the substrate.
Not all fungi can be isolated into pure culture, so
many promising fungi cannot be grown by this method.
Pisolithus tinctorius, Laccaria bicolor (Maire) Orton,
other *Laccaria* spp., some *Hebeloma* spp. and some
Scleroderma spp. have been grown and used effectively.

Prepared vegetative inoculum of some fungi can be
purchased from commercial sources. Mycorr Tech Inc.
(440 William Pitt Way, Pittsburgh, Pennsylvania 15238)
produces vegetative mycelium grown in vermiculite
amended with a proprietary nutrient solution.
Inoculum is produced in 10 liter bags and does not
require leaching of the substrate to remove excess
nutrient solution. This patented process allows
custom production of inoculum of individual fungal
strains to meet research or operational demands.
Various isolates of *P. tinctorius, Laccaria* spp., and
Hebeloma spp. are produced upon request. Recommended
inoculation rate varies for each isolate and fungal
species but is on the order of 1 liter inoculum per
10-12 linear feet of 4-foot nursery bed, or 1 liter
inoculum for 50-100 liters of growth medium in
container nurseries.

In addition, other effective types of vegetative
inoculum have been developed recently. For example,
Mycobeads™ is fungal mycelia immobilized (not
homogenized) in hydrogel beads. These discrete
spherical beads are made in sizes from 0.5 to 3.0 mm
in diameter. Fungal species that have been effective
as this form of inoculum are *Laccaria laccata*
(Scop.:Fr.) Cooke, *Descolea maculata* Bougher,
*Setchelliogaster sp., P. tinctorius and Hebeloma
westraliense* Bougher, Tommerup, & Malajczuk. Inoculum
comes packaged in axenic bags containing water and
beads and can be stored at 4°C for up to 7 months
without loss of propagule quality. Recommended dosage
in sand substrate is 10 beads per seed or 10 beads per
50 cm^3. Mycobeads™ is supplied by Biosynthectica
Ltd., 182 Claisebrook Road, East Perth, Western
Australia 6004.

Another form of vegetative inoculum is mycelium

entrapped in calcium alginate gel (29,34). The fungus
is first grown aseptically in liquid nutrient solution
on a rotary shaker where it forms beads in the media.
The beads are then aseptically washed with
deminerialized water and homogenized in a blender for
5 seconds. Equal volumes of the resulting homogenate
and a 2% w/v sodium alginate solution are mixed with
some added powdered bentonite or milled peat.
Suspensions are then homogenized and passed through
pipette tips so that drops fall into a 0.7 M $CaCl_2$
solution where they form beads that are cured for 45
minutes in the $CaCl_2$ solution.

Spores

Spore inoculum is available as liquid suspensions,
dried pellets or encapsulated seed. Efficacious spore
inoculum is available for some fungal species that are
ineffective as vegetative inoculum such as *Rhizopogon*
spp. In contrast, spore inoculum of *P. tinctorius* is
not as effective as vegetative inoculum.

Spore suspensions of truffle-like fungi such as
Rhizopogon spp. can be prepared as follows.
Sporocarps are gathered, rinsed with tap water to
remove adherent soil and debris, then cut into pieces
(±2 cm^3), and blended with distilled water at high
speed for 2 to 3 minutes, until all pieces are
thoroughly blended to a thick, milky consistency. The
resulting spore concentration is determined with a
hemacytometer and the suspension is stored in the dark
at 5°C until used. Some suspensions can be stored up
to 2 years with only negligible effects on spore
viability and thus inoculum effectiveness. Spores are
applied at about 10,000 spores per seedling at the
time of feeder root production (6-8 weeks after seed
germination). Spores can be delivered to the
seedlings through watering cans or irrigation systems
equipped with a fertilizer injection system. Most
spores of ectomycorrhizal fungi are less than 50μm in
diameter and will pass through most irrigation filters
and nozzle tips. Spores should be applied twice, at a
two-three week interval, to assure even distribution
of inoculum. Spore suspensions of various fungal
species are available from Mycorr Tech (see above for

address) and Forest Mycorrhizal Applications (FMA), 1032 Starlite Dr., Grants Pass, Oregon 97526. Spore suspensions are relatively inexpensive to purchase and rapidly applied compared to vegetative inoculum.

Alternatively, International Forest Seed Co., Box 290, Odenville, Alabama produces spore pellets and seed encapsulated with *Pisolithus tinctorius* spores. Both methods have been effective ways to inoculate various pine species in the eastern United States (27).

The key to a successful mycorrhiza management program is flexibility, adaptability, and clear objectives. Cost of mycorrhiza management can be recovered through increased seedling growth or a reduction in seedling cull percent in the nursery, or increased survival or growth following outplanting. No one fungal species, isolate or inoculum type is appropriate for use worldwide, rather an array of efficacious inocula types of various fungal species exists, depending on nursery cultural practices, host species and economics. For example, Mycobead™ inoculum of *P. tinctorius* and *H. westraliense* is used in Western Australia, spore suspensions of *Rhizopogon* spp. are used in the Pacific Northwest United States and western Europe, vegetative inoculum of *L. bicolor* is used in France and eastern Canada, and vegetative inoculum of *P. tinctorius* is used extensively in the eastern United States and some developing countries.

Pisolithus tinctorius has been the most extensively and intensively studied ectomycorrhizal fungus to date and is used successfully with native pines in the eastern United States and abroad with the introduction of pines and eucalypts throughout the world. In contrast, *P. tinctorius* has been less successful in other regions of North America, specifically the Pacific Northwest and Canada (8).

As additional data is gathered on the biology and ecology of ectomycorrhizal fungi in native forests throughout the world, new fungal isolates and species will have potential for management. In addition, adaptation and simplification of molecular biological techniques affords opportunities to engineer ectomycorrhizal fungi for specific advantages and combine desirable attributes into a customized isolate for use in reforestation programs (1).

FUTURE CONSIDERATIONS

The numerous studies listed in (Table 1), when published, will significantly aid our understanding of the effects of mycorrhizal inoculation on reforestation. Unfortunately, new results are likely to be site specific and will be similar to many reports in the literature. The plants and fungi used are restricted in scope and species; those that are regionally endemic and have potential on a more local basis, are especially lacking.

Forest environments and landscapes are vastly heterogeneous, yet we evaluate the response of a particular fungal isolate-host (seed source) combination through a thin slice of these parameters and then judge the mycorrhizal association on this limited data set. Cline et al. (9) showed different temperature optima for different isolates of various ectomycorrhizal fungi grown in pure culture. Hung and Trappe (18) found similar infraspecific differences in fungal isolate growth in response to pH. Cline and Reid (10) found that ectomycorrhiza formation is influenced by host seed source and fungal species. Lundeberg (21) found that *Pinus sylvestris* L. seedlings planted near to the original seed source formed significantly more ectomycorrhizae than seedlings planted away from the seed collection area. The host seed source x fungal ecotype interaction needs further exploration to understand the physiological adaptations resulting from coevolution. We may find, as Moser (35) suggests, that equivalent provenance's of fungi and tree seed need to be obtained for optimal outplanting success.

One uncontrolled variable in all outplanting trials is the weather. Each outplanting trial is tested under weather conditions that can vary dramatically during the season and from year to year. This changing weather regime complicates response prediction from year to year regardless of other parameters, because the fungus-host combination can respond differently to different soil moistures and temperatures at planting, early season precipitation, summer drought conditions, etc. Unfortunately, it has not been economically feasible to duplicate experiments over time to evaluate this variable.

CONCLUSIONS

Even though the body of literature on mycorrhizae expands at an ever increasing rate, reports on outplanting performance of ectomycorrhiza-inoculated seedling stock remain unfocused and confusing.

The importance of the mycorrhizal habit in establishment and survival of forest trees is not in doubt. Which fungi stimulate particular tree species under specific abiotic and biotic conditions is still an open question. Most biotic and abiotic parameters are not studied rigorously enough to allow coherent predictions across a wide range of conditions.

Considerable research on effects of ectomycorrhizal inoculation on outplanting performance of forest seedlings continues. Unfortunately, most of the current work is similar to previous work using species of Pinaceae as the hosts and *Pisolithus* species as the inoculated mycorrhizal fungus. Additional attention needs to be focused on native plants and host-specific mycorrhizal fungi. Although the applicability of the results are not as wide ranging, the potential for outplanting improvement may well be higher.

TABLE 1. Current (1992) outplanting experiments listed by host species, fungus used, location, and contact.

Host	Fungus	Location[a]	Contact
Acacia difficilis Maiden	*Pisolithus* sp.	N.T., Australia	Reddell
Acacia difficilis	*Scleroderma* sp.	N.T., Australia	Reddell
Acacia gonocarpa F. Muell.	*Pisolithus* sp.	N.T., Australia	Reddell
Acacia gonocarpa	*Scleroderma* sp.	N.T., Australia	Reddell
Acacia holosericea G. Don	*Pisolithus* sp.	N.T., Australia	Reddell
Acacia holosericea	*Scleroderma* sp.	N.T., Australia	Reddell
Acacia mangium Willd.	*Scleroderma* sp.	Queensland	Reddell
Acacia mangium	*Scleroderma* sp.	Vanuatu	Reddell
Acacia platycarpa F. Muell.	*Pisolithus* sp.	N.T., Australia	Reddell
Acacia platycarpa	*Scleroderma* sp.	N.T., Australia	Reddell
Acacia torulosa Benth.	*Pisolithus* sp.	N.T., Australia	Reddell
Acacia torulosa	*Scleroderma* sp.	N.T., Australia	Reddell
Eucalyptus bleeseri Blakely	*Pisolithus* sp.	N.T., Australia	Reddell
E. bleeseri	*Scleroderma* sp.	N.T., Australia	Reddell
E. deglupta Blume	*Pisolithus tinctorius*	(2) Philippines	de la Cruz
E. diversicolor F. Muell.	*Amanita* sp.	Western Australia	Thomson
E. diversicolor	*Protubera canescens* Beaton & Malajczuk	Western Australia	Thomson
E. diversicolor	*Setchelliogaster* sp.	Western Australia	Thomson
E. globulus Labill.	*Amanita* sp.	Western Australia	Thomson
E. globulus	*Descolea* sp.	Western Australia	Thomson
E. globulus	*Protubera canescens*	Western Australia	Thomson
E. globulus	*Setchelliogaster* sp.	(2) Western Australia	Thomson
E. miniata Cunn. ex Schauer	*Pisolithus* sp.	N.T., Australia	Reddell
E. miniata	*Scleroderma* sp.	N.T., Australia	Reddell

Host	Fungus	Location[a]	Contact
E. porrecta S.T. Blake	*Pisolithus* sp.	N.T., Australia	Reddell
E. porrecta	*Scleroderma* sp.	N.T., Australia	Reddell
E. tectifera F. Muell.	*Pisolithus* sp.	N.T., Australia	Reddell
E. tectifera	*Scleroderma* sp.	N.T., Australia	Reddell
E. tetrodonta F. Muell.	*Pisolithus* sp.	N.T., Australia	Reddell
E. tetrodonta	*Scleroderma* sp.	N.T., Australia	Reddell
Larix decidua Mill.	*Laccaria bicolor*	Michigan	Richter
L. decidua	*Scleroderma areolatum* Ehrenb.	Michigan	Richter
L. decidua	*Scleroderma citrinum* Pers.	(2) Michigan	Richter
L. laricina (Du Roi) K. Koch	*Laccaria bicolor*	Quebec, Canada	Gagnon
Picea engelmannii (Parry) Engelm.	*Amphinema byssoides* (Fr.) J. Erikss.	B. C., Canada	Hunt
P. engelmannii	*Complexipes* sp.	B. C., Canada	Hunt
P. engelmannii	*Laccaria laccata*	B. C., Canada	Hunt
P. engelmannii	*Rhizopogon* sp.	B. C., Canada	Hunt
P. glauca (Moench) Voss	*Hebeloma crustuliniforme* (Bull. ex. St. Am.) Quél.	Minnesota	Dixon
P. glauca	*Laccaria bicolor*	Quebec, Canada	Gagnon
P. glauca	*Rhizopogon vinicolor*	Minnesota	Dixon
P. glauca	*Suillus tomentosus* (Kauffm.) Singer, Snell & Dick	Minnesota	Dixon
P. glauca	*Thelephora terrestris* Ehrh. ex Fr.	Quebec, Canada	Gagnon
P. mariana (Mill.) B.S.P.	*Hebeloma crustuliniforme*	N. B., Canada	Salonius
P. mariana	*Hebeloma longicaudum* (Pers. ex Fr.) Kummer	(2) N. S., Canada	Salonius

Host	Fungus	Location[a]	Contact
P. mariana	Hebeloma longicaudum	N. B., Canada	Salonius
P. mariana	Laccaria bicolor	(3) Quebec, Canada	Gagnon
P. sitchensis (Bong.) Carr.	Laccaria proxima Boud.	(4) Scotland	Walker
P. sitchensis	Laccaria proxima	(4) England	Walker
P. sitchensis	Thelephora terrestris	(4) Scotland	Walker
P. sitchensis	Thelephora terrestris	(4) England	Walker
Pinus banksiana Lamb.	Complexipes sp.	Alberta, Canada	Danielson
P. banksiana	Hebeloma crustuliniforme	Minnesota	Dixon
P. banksiana	Hebeloma sp.	Quebec, Canada	Gagnon
P. banksiana	Laccaria bicolor	(8) Quebec, Canada	Gagnon
P. banksiana	Rhizopogon sp.	Quebec, Canada	Gagnon
P. banksiana	Rhizopogon vinicolor	Minnesota	Dixon
P. banksiana	Suillus tomentosus	Minnesota	Dixon
P. banksiana	Tuber sp.	Alberta, Canada	Danielson
P. caribaea Morelet	Pisolithus tinctorius	Thailand	Dixon
P. caribaea	Pisolithus tinctorius	(2) Philippines	de la Cruz
P. clausa (Chapm.) Vasey	Pisolithus tinctorius	(3) Eastern U.S.	Marx
P. contorta Dougl.	Laccaria laccata	(2) B. C., Canada	Hunt
P. echinata Mill.	Pisolithus tinctorius	(3) Eastern U.S.	Marx
P. elliotii Engelm.	Pisolithus tinctorius	(7) Georgia	Marx
P. palustris Mill.	Pisolithus tinctorius	(9) Eastern U.S.	Marx
P. pinaster Ait.	Laccaria bicolor	Spain	Alvarez
P. pinaster	Lyophyllum decastes (Fr.:Fr.) Singer	Spain	Alvarez
P. pinaster	Melanogaster ambiguus (Vitt.) Tul. & Tul.	Spain	Alvarez
P. pinaster	Pisolithus tinctorius	Spain	Alvarez

Host	Fungus	Location[a]	Contact
P. pinaster	*Rhizopogon rubescens* (Tul. & Tul.) Tul. & Tul.	Spain	Alvarez
P. ponderosa Laws.	*Laccaria bicolor*	Oregon	Castellano
P. resinosa Ait.	*Hebeloma crustuliniforme*	Michigan	Richter
P. resinosa	*Laccaria bicolor*	Michigan	Richter
P. resinosa	*Pisolithus tinctorius*	Michigan	Richter
P. resinosa	*Scleroderma areolatum*	Michigan	Richter
P. resinosa	*Scleroderma citrinum*	Minnesota	Dixon
P. resinosa	*Scleroderma citrinum*	Michigan	Richter
P. resinosa	*Scleroderma meridionale* Dem. & Mal.	Michigan	Richter
P. resinosa	*Scleroderma polyrhizum* Pers.	Michigan	Richter
Pinus spp.	*Pisolithus tinctorius*	Jiangxi Prov., China	Marx
P. strobus L.	*Pisolithus tinctorius*	Eastern U.S.	Marx
P. sylvestris	*Hebeloma* spp.	Hebei Prov., China	Marx
P. sylvestris	*Laccaria* spp.	Hebei Prov., China	Marx
P. sylvestris	*Pisolithus tinctorius*	Hebei Prov., China	Marx
P. sylvestris	*Scleroderma citrinum*	Hebei Prov., China	Marx
P. taeda L.	*Pisolithus tinctorius*	(23) Eastern U.S.	Marx
P. virginiana Mill.	*Pisolithus tinctorius*	(2) Eastern U.S.	Marx
Pseudotsuga menziesii	*Hebeloma crustuliniforme*	(2) Oregon	Castellano
P. menziesii	*Laccaria bicolor*	(3) Spain	Alvarez
P. menziesii	*Laccaria bicolor*	England	Walker
P. menziesii	*Laccaria laccata*	(2) B. C., Canada	Hunt
P. menziesii	*Laccaria proxima*	(3) England	Walker
P. menziesii	*Melanogaster ambiguus*	Spain	Alvarez
P. menziesii	*Pisolithus tinctorius*	(2) Oregon	Molina

Host	Fungus	Location[a]	Contact
P. menziesii	*Rhizopogon colossus*	(2) Spain	Alvarez
P. menziesii	*Rhizopogon parksii* Smith	Oregon	Amaranthus
P. menziesii	*Rhizopogon subareolatus* Smith	Spain	Alvarez
P. menziesii	*Rhizopogon villosulus* Zeller	England	Walker
P. menziesii	*Rhizopogon villosulus*	Ireland	Walker
P. menziesii	*Rhizopogon villosulus*	France	Walker
P. menziesii	*Rhizopogon vinicolor*	Spain	Alvarez
P. menziesii	*Rhizopogon vinicolor*	B. C., Canada	Hunt
P. menziesii	*Rhizopogon vinicolor*	(2) Oregon	Amaranthus
P. menziesii	*Rhizopogon vinicolor*	(3) Oregon	Castellano
P. menziesii	*Thelephora terrestris*	(3) England	Walker
Quercus acutissima Carr.	*Pisolithus tinctorius*	Eastern U.S.	Marx
Q. palustris Moench	*Pisolithus tinctorius*	Eastern U.S.	Marx
Tsuga heterophylla (Raf.) Sarg.	*Pisolithus tinctorius*	(2) Oregon	Molina

[a] - Number in parentheses refers to number of plantations with similar parameters.

LITERATURE CITED

1. Barrett, V., Dixon, R.K., and Lemke, P.A. 1990.
    Genetic transformation of a mycorrhizal fungus.
    App. Microbiol. Biotechnol. 33:313-316.
2. Castellano, M.A. 1987. Ectomycorrhizal inoculum
    production and utilization in the Pacific
    Northwestern U.S. - a glimpse at the past, a look
    to the future. Pages 290-292 in: Proceedings
    Seventh North American Conference on Mycorrhizae.
    D.M. Sylvia, L.-L. Hung, and J.H. Graham, eds.,
    Gainesville, FL.
3. Castellano, M.A. 1990. Outplanting performance of
    mycorrhizal inoculated seedlings: a review. Page
    49 in: Proceedings Eighth North American
    Conference on Mycorrhizae. M.F. Allen and S.E.
    Williams, eds., Jackson, WYO.
4. Castellano, M.A. 1993. Outplanting performance of
    mycorrhizal inoculated seedlings: an appraisal.
    New Forests (in press).
5. Castellano, M.A., Luoma, D.L., and Trappe, J.M.
    1990. Preliminary summary of the literature on
    mycorrhiza research. Page 51 in: Proceedings
    Eighth North American Conference on Mycorrhizae.
    M.F. Allen, and S.E. Williams, eds, Jackson, WYO.
6. Castellano, M.A., and Molina, R. 1989.
    Mycorrhizae. Pages 101-167 in: The Container Tree
    Nursery Manual Volume 5. USDA For. Serv. Agric.
    Handb. 674. T.D. Landis, R.W. Tinus, S.E.
    McDonald, and J.P. Barnett, eds., Washington,
    D.C.
7. Castellano, M.A., and Trappe, J.M. 1985.
    Ectomycorrhizal formation and plantation
    performance of Douglas-fir nursery stock
    inoculated with *Rhizopogon* spores. Can. J. For.
    Res. 15:613-617.
8. Castellano, M.A., and Trappe, J.M. 1991.
    Pisolithus tinctorius fails to improve plantation
    performance of inoculated conifers in
    southwestern Oregon. New For. 5:349-358.
9. Cline, M.L., France, R.C., and Reid, C.P.P. 1987.
    Infraspecific and interspecific growth variation
    of ectomycorrhizal fungi at different

temperatures. Can. J. Bot. 65:869-875.

10. Cline, M.L., and Reid, C.P.P. 1982. Seed source and mycorrhizal fungus effects on growth of containerized *Pinus contorta* and *Pinus ponderosa* seedlings. For. Sci. 28:237-250.

11. Garbaye, J., and Bowen, G.D. 1987. Effect of different microflora on the success of ectomycorrhizal inoculation of *Pinus radiata*. Can. J. For. Res. 17:941-943.

12. Garbaye, J., and Bowen, G.D. 1989. Stimulation of ectomycorrhizal infection of *Pinus radiata* by some microorganisms associated with the mantle of ectomycorrhizas. New Phytol. 112:383-388.

13. Garbaye, J., Delwaulle, J.C., and Diangana, D. 1988. Growth response of eucalypts in the Congo to ectomycorrhizal inoculation. For. Ecol. & Mgmnt. 24:151-157.

14. Frank, A.B. 1885. Ueber die auf Wurzelsymbiose beruhende Ernährung gewisser Baüme durch unterirdische Pilze. Ber. Deut. Bot. Ges. 3:128-145.

15. Kamienski, F. 1882. Les organes végétatifs du *Monotropa hypopitys*. Mem. Soc. Nat. Sci. Math. Cherbourg 24:5-40.

16. Kamienski, F. 1985. The vegetative organs of *Monotropa hypopitys* L. Pages 12-17 in: Proceedings Sixth North American Conference on Mycorrhizae. R. Molina, ed., Bend, OR.

17. Kessell, S.L. 1927. Soil organisms. The dependence of certain pine species on a biological soil factor. Empire Forestry 6:70-74.

18. Hung, L.-L., and Trappe, J.M. 1983. Growth variation between and within species of ectomycorrhizal fungi in response to pH *in vitro*. Mycologia 75:234-241.

19. Le Tacon, F., Garbaye, J., Bouchard, D., Chevalier, G., Olivier, J.M., Guimberteau, J., Poitou, N., and Frochot, H. 1988. Field results from ectomycorrhizal inoculation in France. Pages 51-74 in: Proceedings Canadian Workshop on Mycorrhizae in Forestry. M. Lalonde, and Y. Piché, eds., Québec, Canada.

20. Linderman, R.G. 1985. Microbial interactions in

the mycorrhizosphere. Pages 117-120 in: Proceedings Sixth North American Conference on Mycorrhizae. R. Molina, ed., Bend, OR.

21. Lundeberg, G. 1968. The formation of mycorrhizae in different provenances of pine (*Pinus sylvestris* L.). Svensk Bot. Tidskr. 62:249-255.

22. Marx, D.H. 1976. Use of specific mycorrhizal fungi on tree roots for reforestation of disturbed lands. Pages 47-65 in: Proceedings Conference on Forestation of Disturbed Surface Areas; 1976 April: Birmingham, Alabama: USDA-Forest Service, Southeastern Area, State and Private Forestry.

23. Marx, D.H. 1980. Ectomycorrhizal fungus inoculations: a tool for improving forestation practices. Pages 13-71 in: Tropical Mycorrhizae. P. Mikola, ed., Clarendon Press, Oxford.

24. Marx, D.H. 1991. Forest application of the ectomycorrhizal fungus *Pisolithus tinctorius*. Marcus Wallenberg Prize lecture. Stockholm, Sweden. 35 pp.

25. Marx, D.H., Cordell, C.E., Kenney, D.S., Mexal, J.C., Artman, J.D., Riffle, J.W., and Molina, R.J. 1984. Commercial vegetative inoculum of *Pisolithus tinctorius* and inoculation techniques for development of ectomycorrhizae on bare-root seedlings. For. Sci. Monogr. 25:1-101.

26. Marx, D.H., Cordell, C.E., Maul, S.B., and Ruehle, J.L. 1889. Ectomycorrhizal development on pine by *Pisolithus tinctorius* in bare-root and container nurseries. I. Efficacy of various vegetative inoculum formulations. New For. 3:45-56.

27. Marx, D.H., Jarl, K., Ruehle, J.L., and Bell, W. 1984. Development of *Pisolithus tinctorius* ectomycorrhizae on pine seedlings using basidiospore-encapsulated seeds. For. Sci. 30:897-907.

28. Marx, D.H., and Kenney, D.S. 1982. Production of ectomycorrhizal fungus inoculum. Pages 131-146 in: Methods and Principles of Mycorrhizal Research. N.C. Schenck, ed., Amer. Phytopathol. Soc., St. Paul, MN.

29. Mauperin, Ch., Mortier, F., Garbaye, J., Le Tacon, F., and Carr, G. 1987. Viability of an ectomycorrhizal inoculum produced in a liquid medium and entrapped in a calcium alginate gel. Can. J. Bot. 65:2326-2329.
30. Mikola, P. 1970. Mycorrhizal inoculation in afforestation. Int. Rev. For. Res. 3:123-196.
31. Mikola, P. 1973. Application of mycorrhizal symbiosis in forestry practice. Pages 383-411 in: Ectomycorrhizae: their ecology and physiology. G.C. Marks, and T.T. Kozlowski, eds., Academic Press, New York.
32. Molina, R., and Palmer, J.G. 1982. Isolation maintenance, and pure culture manipulation of ectomycorrhizal fungi. Pages 115-129 in: Methods and Principles of Mycorrhizal Research. N.C. Schenck, ed., Amer. Phytopathol. Soc., St. Paul, MN.
33. Molina, R., and Trappe J.M. 1984. Mycorrhiza management in bareroot nurseries. Pages 211-223 in: Forest Nursery Manual: production of bareroot seedlings. M.L. Duryea and T.D. Landis, eds., Martinus Nijhoff/Dr. W. Junk Publ., The Hague.
34. Mortier, F., Le Tacon, F., and Garbaye, J. 1988. Effect of inoculum type and inoculation dose on ectomycorrhizal development, root necrosis and growth of Douglas-fir seedlings inoculated with *Laccaria laccata* in a nursery. Ann. Sci. For. 45:301-310.
35. Moser, M. 1958. Der Einfluss tiefer Temperaturen auf das Wachstum und die Lebenstätigkeit hoherer Pilze mit spezieller Berücksichtigung von Mycorrhizapilzen. Sydowia 12:386-399.
36. O'Dell, T.E., Castellano, M.A., and Trappe, J.M. 1992. Biology and application of ectomycorrhizal fungi. Pages 379-416 in: Soil Microbial Ecology. F.B. Metting, Jr., ed., Marcel Dekker, New York.
37. Riffle, J.W. 1989. Field performance of ponderosa, scots, and Austrian pines with *Pisolithus tinctorius* ectomycorrhizae in prairie soils. For. Sci. 35:935-945.
38. Ruehle, J.L. 1982. Mycorrhizal inoculation improves performance of container-grown pines

planted on adverse sites. Pages 133-135 in: Proceedings Southern Containerized Forest Tree Seedling Conference; 1981 August: New Orleans, Louisiana: USDA-Forest Service, Southern Forest Experiment Station.

39. Schramm, J.R. 1966. Plant colonization studies on black wastes from anthracite mining in Pennsylvania. Trans. Amer. Philosoph. Soc. 56:1-194.

40. Trappe, J.M., and Castellano, M.A. 1992. Mycolit: a mycorrhiza bibliography, 1758-1991. Mycologue Publications, Waterloo, Canada. 550 pp.

41. Vozzo, J.A., and Hacskaylo, E. 1971. Inoculation of *Pinus caribaea* with ectomycorrhizal fungi in Puerto Rico. For. Sci. 17:239-245.

42. Walker, C. 1992. Inoculating Douglas-fir seedlings with mycorrhizal fungi. Forestry Commission, Edinburgh, Scotland, Res. Info. Note 222, 5 pp.

SYSTEMATICS OF GLOMALEAN ENDOMYCORRHIZAL FUNGI: CURRENT VIEWS AND FUTURE DIRECTIONS

Stephen P. Bentivenga and Joseph B. Morton
West Virginia University
Morgantown, West Virginia

The method then that we must adopt is to attempt to recognize the natural groups - each of which combines a multitude of differentiae, and is not defined by a single one as in dichotomy.
— Aristotle, *De Partibus*

In this symposium, the current state of mycorrhizal research in a wide variety of areas ranging from biological control to atmospheric pollution was discussed. All of the topics concern interactions in the ecological hierarchy at the population, community, or ecosystem levels. These interactions depend on the biological organisms comprising the ecological hierarchy. Organisms are also arranged hierarchically in taxonomies based on whatever kinds of diversity exist as stable patterns. This hierarchy traditionally is ordered into formal classifications that serve either as a practical basis for identification of populations or as a theoretical framework for understanding phylogenetic relationships among taxonomic units. Both purposes are important because the former allows recognition of individual organisms in named groups and the latter provides hypotheses of historical patterns of diversity and causal mechanisms. Classifications that serve both purposes equally well are very difficult to construct. To correctly define biological units of comparisons and then interpret results from those comparisons, the basis for a classification must be thoroughly understood.

All classifications are composed of the same Linnaean taxonomic categories (*e.g.*, species, genera, *etc.*). However, the underlying kinds and causes of diversity are less universal, and they often differ greatly within each group of organisms (25). Traditionally, first attempts at classifying organisms are based on convenience groups because data on diversity are scarce. Primary efforts are devoted to description and nomenclature and production of keys for identification. This "alpha taxonomy" (24) provides only an operational database. It offers no information on phylogenetic patterns and processes responsible for current taxonomic groups. This is the realm of systematic analysis (47).

Fungi that form symbiotic arbuscular endomycorrhizal associations pose a unique challenge to anyone interested in taxonomy or systematics. The life history of these fungi consists of two distinct and quasi-independent modes, one somatic and the other reproductive (31,35). The somatic mode encompasses growth and differentiation of vegetative hyphae, arbuscules, and either intraradical vesicles or extraradical auxiliary cells. These parts are implicated in formation and maintenance of the mycorrhizal association in plant roots. Reproductive spores function to package genetic information of the organism, disperse this information, and then initiate new individuals spatially, and potentially temporally, distant from the parent. Spores become physically and physiologically isolated from the vegetative soma early in differentiation and thus do not contribute to activities of mycorrhizal maintenance or host-fungus interactions. Although somatic and reproductive modes are interdependent in maximizing fungal fitness (however defined), asymmetry in form and function results in different rates and magnitudes of morphological divergence (31). Spores were the sole focus of attention of the early taxonomy, which was not formalized using traditional nomenclature until 1974 (12). Taxonomic research still includes a significant exploration and description component (32) because so few areas of the world have been extensively sampled for indigenous species. Current

problems are circular in nature. The first attempt at
systematic analyses (30) and evolutionary explanations
based on cladistic patterns of diversity (31) depended
on faulty interpretations of existing species and
character concepts. However, this database can be
improved only by a strong theoretical base developed
from thorough systematic studies.
     We are at a critical juncture in the transition
from a classification based on alpha taxonomy to one
derived from systematic studies. In this chapter, we
will discuss the state of taxonomic and systematic
research for endomycorrhizal fungi, encompassing both
problem areas and some proposed solutions. Finally,
we will examine the implications of these studies to
researchers and outline important future directions.

Alpha Taxonomy and Identification

     The approach to classifying endomycorrhizal fungi
in the 1960's consisted of lumping all species into
the genus *Endogone sensu lato* of the Zygomycetes.
Grouping criteria included an ecological component in
that all species were thought to occupy a similar
niche (associations with roots of higher plants). A
landmark in classification was the scheme proposed by
Gerdemann and Trappe (12), who recognized the genera
Acaulospora, *Gigaspora*, *Glaziella*, *Modicella*, and
Sclerocystis together with *Endogone* in Endogonaceae,
Zygomycetes. Their classification was phyletic, since
it was based on *a priori* estimates of relationship
based on spore organization. Few characters were
known at the time, so their stability and historical
significance were only approximations. As more
species were described, new characters were discovered
and the information base grew. *Glaziella* and
Modicella were transferred to the Ascomycetes (13) and
Mortierellaceae (Zygomycetes) (51), respectively.
Ames and Schneider erected a new genus, *Entrophospora*,
and Walker and Sanders (59) placed species of
Gigaspora sensu lato with inner walls and a
germination shield into a new genus, *Scutellospora*.
To sort out diversity in phenotypically distinct
microscopic structures within spores of the
endomycorrhizal fungi, Walker (54,55) and then others

(5,28,46) defined phenotypically different and separable wall types. A graphical representation (called a murograph) of the types and relative position of "walls" was developed to permit more concise and standardized descriptions and comparisons of spores belonging to different species. New species described with murographs continued to be forced into the artificial classification of Gerdemann and Trappe (12) because no other rationale was formally proposed. Pirozynski and Dalpé (39), realizing the tenuous relationship between *Glomus* and *Endogone*, separated *Glomus* and *Sclerocystis* into their own family, Glomaceae. This was followed by a revision of the Endogonales *sensu lato* by Morton and Benny (33), who partitioned all arbuscule-forming fungi into one order, Glomales. Members of Glomales were separated further into two suborders (Glomineae and Gigasporineae) and three families (Acaulosporaceae, Gigasporaceae, Glomaceae) based on morphological characters of fungal soma and spores (Table 1). Subsequently, all species in the genus *Sclerocystis*, except *S. coremioides* Berk. & Broome, were transferred to *Glomus* (2).

Currently, 149 species of glomalean fungi have been described and named according to guidelines set forth in the International Code of Botanical Nomenclature (14). Recent changes in the nomenclature of species (1,60) and genera (57) are trivial with no loss of important information. They represent important housekeeping chores, but they do not address crucial goals: to understand the origin and causation of diversity in glomalean fungi and to organize that diversity into biologically meaningful taxa.

More critical nomenclatorial problems concern type specimens. The type method, as currently practiced, prevents any correspondence between the morphology of spores observed at the time the species name was conceived and after several months in storage media (29,32). Thus, characters in many types are not likely to correspond to those characters present in freshly extracted spores. If identity between original specimens and the species name are to be preserved at all, then supplemental procedures must be

Table 1. Summary of taxonomic characters used to classify fungi in Glomales, Zygomycetes, using wall terminology first proposed by Walker (54).

Order: Glomales Morton & Benny
   Soil-born fungi characterized by transient dichotomously-branched arbuscules in cortical cells of plant roots after establishing an obligate mutualistic symbiosis with many plant species. Spores appear to be obligately asexual, forming within or outside roots. Most species diversity is manifested in spore size, color, and microscopic features of subcellular spore walls distinct in structural properties and histochemical properties (usually Melzer's reagent).
   Suborder: Glomineae Morton & Benny
      Arbuscular fungi forming intraradical vesicles in mycorrhizal roots; spores born terminally, intercalarily, and laterally from one or more subtending hyphae. Members are referred to as "vesicular-arbuscular mycorrhizal (VAM) fungi".
      Family: Glomaceae Pirozynski & Dalpe'
         Spores (i) are produced singly, in aggregates, in an unorganized hyphal matrix, or in a highly ordered hyphal matrix, with the structural wall continuous with a wall of the subtending hypha; (ii) show morphological diversification mostly in the number and types of walls formed outside the structural wall (e.g., evanescent, expanding, mucilagenous, and unit walls); (iii) form flexible inner walls so far limited to those which are membranous, usually one or two in number, and which rarely stain positive in Melzer's reagent; (iv) seal off contents from that of the subtending hypha by different mechanisms, such as an amorphous plug, a septum, an inner membranous wall, or thickening of the structural wall; and (v) germinate usually by emergence of the germ tube through the subtending hypha.
         Genus: Glomus Tulasne & Tulasne
            Has all familial characters, except that spores are not formed in a highly organized matrix originating from a columnar base.
         Genus: Sclerocystis Berkeley & Broome emend. Almeida & Schenck
            Chlamydospores formed around a central hyphal plexus originating from a columnar hyphae base. Only one species: S. coremioides Berk. & Broome.
      Family: Acaulosporaceae Morton & Benny
         Spores (i) are borne laterally from or within the neck of a sporiferous saccule that is formed terminally on a fertile hyphae; usually singly but occasionally in aggregates; (ii) have an outer wall which is continuous with the subtending hyphal wall and which sloughs with age; (iii) are sessile following extraction from soil because of sloughing of attached hyphae, (iv) have a smooth to highly ornamented structural wall that does not originate from the wall of the subtending hypha; (v) form at least one flexible inner wall, but usually two or more; (vi) show most diversification in number and types of inner walls (e.g., the semi-rigid unit wall, beaded or smooth membranous wall, and amorphous wall), with the innermost wall types often producing a dextrinoid to dark red-purple reaction in Melzer's reagent; (vii) seal off contents from that of the subtending saccule neck by a plug indistinguishable from the structural

Table 1 (continued)

wall; and (viii) germinate from a germ tube emerging from a spherical "germination shield" which forms between a semi-rigid unit wall and the innermost pair of inner flexible walls when such walls are synthesized.

<u>Genus</u>: *Acaulospora* Gerdemann & Trappe emend. Berch
Spores borne laterally on the neck of the sporiferous saccule in a continuous transition series from those with a glomus-like hyphal attachment to those borne on a pedicel and finally to those borne on a short collar.

<u>Genus</u>: *Entrophospora* Ames & Schneider
Spores are formed within the neck of the sporiferous saccule. Spore ontogenesis and spore wall diversity mirror that in *Acaulospora*.

<u>Suborder</u>: <u>Gigasporineae</u> Morton & Benny
Arbuscular fungi which also form extraradical auxiliary cells singly or in clusters; globose to subglobose spores often exceeding 200 μm in diameter, forming on a bulbous sporogenous cell. Members of the group do not form vesicles and hence are "arbuscular mycorrhizal fungi".

Family: Gigasporaceae Gerd. & Trappe
Spores (i) always have a persistent thin outer wall which encloses and adheres to a structural laminated wall and (ii) seal off contents from that of the bulbous sporogenous cell by a thin septum or plug.

<u>Genus</u>: *Gigaspora* Gerd. & Trappe emend. Walker & Sanders
Spores do not differentiate any flexible inner walls; germ tubes arise from a warty germinal wall which rarely separates from the laminated wall; auxiliary cells usually echinulate.

<u>Genus</u>: *Scutellospora* Walker & Sanders
Spores (i) always differentiate two or more flexible inner walls which often form in adherent pairs (ii) show most diversification in the number and types of inner walls (e.g., membranous walls of different thickness, coriaceous wall, amorphous wall), with the innermost flexible wall often producing a dextrinoid to dark red-purple reaction in Melzer's reagent, (iii) germinate via germ tubes arising from a persistent germination shield of variable shape and margin which always forms on the innermost pair of flexible inner walls. Auxiliary cells smooth to knobby.

---

implemented. These include propagation of fungi followed by cryopreservation and more extensive and detailed photographs of specimens. Currently, morphological characters on and within spores provide the basis for species delimitation and identification (Table 1; Figure 1). This task is not easy for inexperienced workers because rounded and smooth soil crystalline deposits, nematode and insect eggs, sclerotia, algal cells, and other structures often are mistaken for spores in extracts from both field and pot culture soils. Subcellular characters such as the various kinds of "walls" proposed by Walker (54) and others (29) and wall properties such as thickness, pigmentation, ornamentations, and histochemical reactions can only be examined in broken spores. Most "walls" are difficult to distinguish as individual characters because they are colorless (unless immersed in Melzer's or other histochemical reagents) and they tend to wrinkle, fold, and overlap. Some "walls" become separated easily, while others remain attached regardless of manipulations. Experience at recognizing these characters comes from examination of healthy spores of a representative range of species.

Subcellular organization and structure of spores is assumed to be unaffected by the environment. In reality, field-collected spores often are either parasitized, degraded, or modified enough in subtle ways to lead to misinterpretations of subcellular characters (29,32). Thus, field-collected spores cannot be assumed to possess all of the informative characters for species-level identification. The only direct solution to these problems is to set up "bait" or "trap" cultures of indigenous fungi in a greenhouse. The procedure is simple: rhizosphere soil is mixed with a coarse sand, seeded with a host plant of the same or related species as that at the source location, and left to grow for 3-4 months. Spores extracted from bait cultures are much more uniform in their morphologies, and are abundant enough to provide a large sample size to examine the full range of morphological variation in spores. Most trap cultures contain a mixture of species, and these can be established singly in culture using spores inoculated directly onto roots of a suitable host.

Inadequate knowledge of the origin and nature of characters and their role in defining a species taxon accounts for many errors in published species descriptions (32). Some species have been described from parasitized (7,36) or preserved (6) specimens. Published keys and manuscripts based solely on species descriptions (4,15,50) perpetuate these errors. Researchers are justifiably frustrated by this situation, but it is a function of the young age of glomalean taxonomy. The solution is to extract,

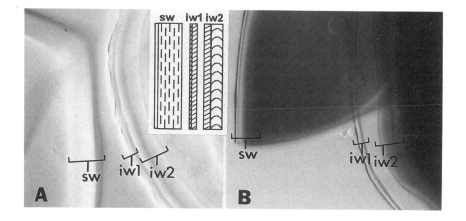

Figure 1. Subcellular structures in spores of
Scutellospora pellucida (with murograph) as
distinguished by unique origin in differentiation. sw
= spore wall; iw1 = first inner wall; iw2 = second
inner wall. A.) Crushed spore in polyvinyl lactic
acid glycerol (PVLG). Differential interference
contrast microscopy (DIC), × 780. B.) Crushed spore
in PVLG with Melzer's reagent. DIC, × 750. Murograph:
no fill = unit layer; dashed fill = laminae; diagonal
fill = flexible layer; semi-circle fill = amorphous
layer.

mount, and study whole and broken spores of healthy
specimens from pot cultures. With this approach, or
even when spores are collected from field material,
identification must be made at least to genus. The
short-term solution to species identification is to
prepare a set of voucher specimens (slides, vials of
intact spores, photographs) of each fungus. If the
research has important implications or is ever to be
repeated, then permanent physical evidence must exist
for comparative purposes.

The International Culture Collection of Arbuscular
and Vesicular-arbuscular Mycorrhizal Fungi (INVAM) is
an important resource for scientists who do not have
the time or facilities to carry out these activities.
INVAM was set up as a germplasm resource. In addition

to maintaining over 650 isolates of 76 species (at least 20 of which are undescribed), other services are offered to facilitate taxonomic decisions. The collection includes over 2000 vouchers of slides with permanently mounted spores, over 600 vouchers of vials containing preserved spores, and over 1200 color photographs of broken spores and their diagnostic features. A newsletter is mailed to subscribers twice each year to communicate up-to-date information on culture availability, tips and problems associated with culture methods, new taxonomic information or reevaluations of existing taxa, *etc.* All of these materials are available for researchers to purchase or borrow. One avenue researchers can take to identify their fungal isolates is to contribute them to INVAM where they must be identified at the time they are numbered as new accessions.

Systematics and Phylogeny

Questions of how evolutionary change occurred in glomalean fungi and the nature and magnitude of these changes need to be answered if mycorrhizal processes are to be fully understood. However, they can be addressed only from hypotheses of phylogenetic relationships generated by cladistic methods (47 and 48 for comparison of these with phenetic and phyletic approaches). Results of phylogenetic studies are not often incorporated into the structure of many classifications because entrenched taxonomies would be destabilized enough to limit their utility for other purposes such as identification. However, when classifications are based on phylogenies, they provide an invaluable tool in testing relationships using other character data sets (genetic, biochemical, physiological, ultrastructural, biogeographic, ecological). Such is the case in the transition of glomalean classification to reflect phylogenetic relationships (33). Despite some inherent flaws in the approaches used and a weak character data base (32,34), the major patterns of evolutionary change have begun to provide new insights into micro- and macroevolutionary processes (31).

Characters comprise the data used in phenetic,

cladistic, or other less acceptable methods for phylogeny reconstruction (47). Kohn (19) describes a wide range of characters potentially useful for fungal systematics as well as methods for determining their level of resolution. Taxonomically informative characters can be molecular/biochemical, biogeographical, ecological, physiological, as well as morphological. In any phylogenetic analysis, however, not just any characters suffice. Only homologous characters estimate phylogenetic relationships (9,63). Homology is hypothesized when stable characters shared by different taxa are similar enough to signify common ancestry. Tests of similarity include equivalency in: (a) composition and structure, (b) adjacent structures, and (c) ontogenetic sequences (38,53). Characters which fail these tests are rejected as homologs and then are interpreted as analogous (having independent ancestry) (41). Cladistic analysis includes a time component because ancestral (primitive) homologs precede descendant (derived) characters when distributed as internested sets in cladograms (63). Only derived homologs define branch points, so they must be identified through the process of polarization. Two methods are used to make polarity decisions: outgroup analysis (61) or the ontogenetic method (18), with the former having more universal applicability. When neither method yields enough information on characters, those characters cannot be included in phylogenetic analysis.

From the above discussion, it should be apparent that character selection depends on rigorous comparative analysis to determine: (a) stability, (b) homology, and (c) evolutionary directionality, or polarity. Character stability can be measured directly by placing fungi in different environments (27) and indirectly by determining the distribution of a similar character in different taxa (30). Problems arise when phylogenetic affinities are presumed from incomplete evidence. Consider that similarities in wound-healing of *Acaulospora nicolsonii* Walker, Reed & Sanders and *Gigaspora gigantea* (Nicol. & Gerd.) Gerdemann & Trappe were proposed as evidence for close phylogenetic ties of *Acaulospora* and *Gigaspora* (58), yet they involved a comparison of only two species. Similarly, Jabaji-Hare (17) concluded that glomalean fungi are closely related to chytrids and hyphochytrids based on the presence of particular fatty acids without taking into account the plasticity of these characters relative to substrate specificity (21).

Homology also can be assessed directly from ontogenetic studies if characters appear in a linear pattern during developmental processes (18,41). In spores of glomalean fungi, subcellular structures meet this requirement by differentiating in a linear sequence of discrete stages distinguishable either by appearance of new characters or transformations of characters from one morphological state to another (11,32,34,35).

The phenotypically distinct and separable "walls" defined by Walker (54,55), Berch and Koske (5), Morton (28), and Spain *et al.* (46) have been treated diagnostically and conceptually in glomalean taxonomy as separate and independent structures (*e.g.*, six in the case of *S. pellucida*, Figure 1). While this view had important and beneficial utility in alpha taxonomy, the causal basis for discrete differences in mature spore phenotypes has never been addressed. Berch (3) argues that glomalean fungi possess only one spore wall with many layers, with the assumption that all subcellular diversity arose from a common origin. Unfortunately, her view also was devoid of hard experimental evidence from developmental comparisons.

Developmental evidence now is beginning to show that spore walls (*sensu* Walker) are not separate characters of independent origin (11,34). In the most complex spores such as those of *S. pellucida*, for example, only three walls of separate origin are differentiated (Figure 1). They correspond in most species to what previously were called wall groups (29,54). Within each wall are distinct layers of common origin (thus belonging to the same structure) which differentiate qualitative and quantitative characters such as color or thickness, respectively. The spore wall can be defined developmentally as the only subcellular structure to differentiate concomitantly with spore expansion. Once the spore attains its final size and the spore wall has terminated differentiation, then one or more flexible inner walls are formed sequentially in some groups (*Acaulospora*, *Entrophospora*, *Scutellospora*, possibly *Glomus*). These inner walls are linked functionally to germination processes.

Differentiation patterns of subcellular structures in a hierarchical order from primary to tertiary characters (Figure 2) provide conclusive evidence that characters do not have equal taxonomic (or phylogenetic) value or weight. Some are so conserved that they are shared (inherited by several descendants from a common ancestor), so they define higher taxonomic categories. Others are more changeable and delimit only single species. These differences require well-defined weighting procedures, which have yet to be established from developmental patterns. Primary characters such as the three walls of *S. pellucida* (Figure 1), are likely to receive

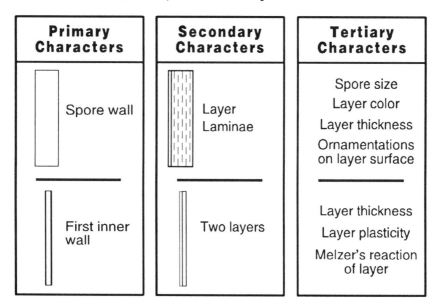

Primary Characters	Secondary Characters	Tertiary Characters
Spore wall	Layer Laminae	Spore size Layer color Layer thickness Ornamentations on layer surface
First inner wall	Two layers	Layer thickness Layer plasticity Melzer's reaction of layer

Figure 2. Heirarchical depiction of taxonomic characters in *Scutellospora pellucida* (without the second inner wall). See text for explanation.

the most weight because they arise independently
during spore differentiation and are shared by members
of families and suborders as currently circumscribed
(33, Table 1). Secondary characters are the
phenotypically distinct layers of primary structures
(Figure 2). They currently circumscribe the genus
Gigaspora, but equivalent groups exist in other genera
that currently are not recognized in any taxonomic
scheme (Morton, unpublished). Tertiary characters are
morphological differences within each layer, such as
color variation, layer thickness, and ornamentation
patterns (Figure 2). They define most species in all
genera of Glomales.

The weighting described above is based causally on
degree of developmental constraints on characters
(their conservation and distribution among taxa).
Therefore, the procedure is tied to real biological
mechanisms associated with developmental (or
epigenetic) phenomena. As both the patterns of
character evolution and the constraints imposed on
character variation become better understood, then
identification of existing taxa will become easier and
predictions of expected diversity in undiscovered
species will become possible. For example, many
species of *Acaulospora* differ only in ornamentations
of spore wall layers, while inner walls are identical
(Morton, unpublished). Thus, new species are expected
which vary only in kinds of ornamentations. In this
case, problems of seeing fine, minute details of inner
walls becomes less important, and the focus shifts to
aspects of the spore wall which are more easily
interpreted under the microscope.

Ecological characters are those that lead to
establishment of a functional symbiosis and to
adaptation of an organism to its local environment.
By their nature they are plastic and reversible, and
thus unlikely to contribute to delimiting taxonomic
groups (34). However, it is these characters that are
of concern to ecologists and agronomists interested in
fungal contributions to plant growth. Thus, the
contention that a better understanding of taxonomic
characters will lead to increased prediction of fungal
physiological behavior (56) is not valid.

Molecular and biochemical characters have been
examined only recently in glomalean fungi. They
include cell wall composition (62), protein
immunogenicity (64,43), isozyme electrophoretic
patterns (42), fatty acid methyl ester (FAME) profiles
(17), and direct sequencing of the 18S rRNA gene (44).
Researchers have been hampered by the inability to
culture arbuscular fungi and the difficulty in
extracting and purifying large amounts of DNA from
spores (10). However, the polymerase chain reaction
(PCR) technology now provides the means to amplify DNA

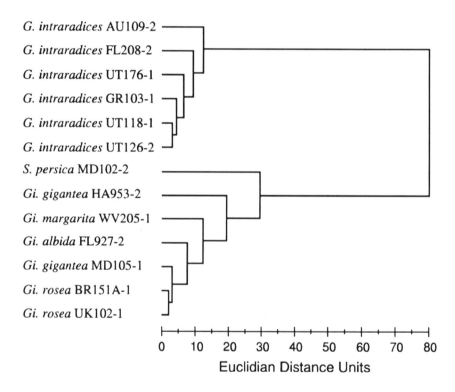

**Figure 3. Dendrogram of overall similarity (expressed
as Euclidian distance) of fatty acid methyl ester
(FAME) profiles of isolates of** *Glomus intraradices*,
Scutellospora persica, **and** *Gigaspora* **spp.**

from single spores (Millner, pers. comm.). Other
characters may be obtained from analysis of DNA
restriction fragment length polymorphisms (RFLP's),
nuclear karyotyping (either microscopically or
indirectly using pulse-field gel electrophoresis) and
DNA-DNA hybridization. Presently, molecular
characters are useful only in inferring phenetic
relationships (disguised as phylogenetic analyses)
because polarity and thus ancestor-descendent
relationships are difficult to determine. Numerical
methods (45) are most useful in defining diagnostic
characters or circumscribing clusters of taxa based on
estimates of overall similarity. Figure 3 illustrates
the relative similarity (expressed as Euclidian
distance) in the FAME profiles of isolates of *Glomus
intraradices* and several members of Gigasporaceae.
Other preliminary results (Bentivenga and Morton,
unpublished) also suggest that individual fatty acids
are extremely stable and thus provide characters for
incorporation into a cladistic analysis.

Implications of Systematic Patterns

Species are the basic experimental unit in a
cladistic analysis and basic taxonomic unit in a
classification. They clearly exist as coherent
entities uniting populations of glomalean fungi (31).
The most fundamental questions concerning these and
any other species are: What mechanism accounts for
species coherence? How do species participate in
present day ecological interactions? Many species
concepts have been forwarded based on population,
ecological, or evolutionary processes, which differ a
great deal among groups or organisms (26). Glomalean
species cannot be based on population phenomena
because evidence for sexual reproduction is rare (see
49 for the only example), and evidence for gene flow
is nonexistent. Species based on ecological
boundaries are defined by co-occupation of a similar
adaptive zone (52). However, all glomalean fungi
occupy a similar collective niche (31), and taxonomic
groups do not correlate with environmental parameters
or gradients. Species also may be delimited by
phylogenetic relationships (25). The phylogeny

proposed by Morton and Benny (30) rejects the notion that glomalean taxa are artificial agglomerations of independent species (with independent characters). Instead, species are linked as a network of ancestors and descendants. Only conserved characters passed from a unique ancestor to all descendants unite all asexual populations of a glomalean fungus into a species. The mechanism for conservation in form is known: constraints imposed during spore differentiation. With this knowledge, Morton *et al.* (35) defined species of glomalean fungi as:

> *the smallest assemblage of reproductively isolated individuals or populations diagnosed by epigenetic, morphological or organizational properties of fungal spores that specify a unique genealogical origin and continuity of descent according to the criterion of monophyly.*

Some terminology in this definition may be unfamiliar to some scientists, but it most succinctly establishes the biological basis of the discontinuities that delimit species and the phylogenetic criteria for grouping and ranking populations into a species. Grouping is based on homologies that establish a common ancestor for all populations of a species (e.g. monophyly). Ranking is based on the smallest diagnosable group with these homologs. Operationally, homologs are hypothesized first by passing rigorous tests of similarity and verified by cladistic analysis. In compositionally and phenotypically simple characters within spores that have limited avenues for divergence, convergent evolution is a very real possibility that can only be resolved by these dual analyses. This does not mean that alpha taxonomy cannot proceed as usual. Because this species concept is a theoretical construct, it will supplement, rather than conflict with, normal operations and decisions made in matters of alpha taxonomy (see above). More importantly, the concept helps researchers to understand the nature of glomalean species, how they compare with other living organisms, and when and how they can serve as experimental units in research endeavors.

Most speculation on the evolution of the endomycorrhizal association (20,22,23,37,39) involves evolution of only the somatic portion of the fungus and thus has taxonomic relevance only at the family level or above where morphological characters (arbuscules, vesicles, hyphal patterns) are linked to mycorrhizal functions. But morphological speciation in glomalean fungi appears to be confined to diversification of subcellular characters in spores. If developmental constraints on variation in spore characters is as great as ontogenetic comparisons indicate (11,34), then physiological and genetic evolution has and continues to occur independent of stable spore morphologies. Therefore, speculation about evolution of the mycorrhiza does not provide much insight into evolution of the spore phenotype or vice versa.

Taxonomic diversity often is interpreted as much by rate and degree of extinction as it is by the magnitude of speciation events (8). Extinctions are advantageous in providing convenient gaps between taxonomic groups. Glomalean fungi have tremendous potential for longevity because of their life history traits (31). Since the fungal soma and its reproductive spores are totipotent, any dispersal insures that some descendant parts survive even if most or all of an organism at a given location dies. We predict that the result of this will be coexistence of ancestral and descendant phenotypes resulting from morphological evolution. The result will be very small gaps between species that will reside in tertiary characters (Figure 2). This means that emphasis must be placed on defining these intergrading characters in a way that establishes their stability in uniting different geographic populations into a species taxon.

Much of the previously published mycorrhizal literature treats the species as the taxonomic unit participating in the symbiosis. This view stems from the erroneous conclusion that arbuscular mycorrhizal fungi behave similarly to animals and plants that possess many functionally important taxonomic characters. Differentiation and function of important taxonomic characters of spores are completely uncoupled from those of the mycorrhiza. Thus, the participatory unit in the mycorrhizal symbiosis is the individual organism either alone or in populations rather than the species. This concept, while long a basic tenet of plant pathological interactions, is only now beginning to gain acceptance among mycorrhizologists.

Future Directions

Alpha taxonomy, despite its traditional approaches, serves an essential function for identification purposes and building a diversity database. Therefore, urgent problems must be resolved with haste. For the time being, descriptions of known species must be revised and this task should receive a higher priority than descriptions of new species. Only with this information can rational keys can be devised for accurate and consistent identifications. Since these revisions will be accurate *only* when based on *freshly-extracted* spores propagated in pot or other plant-associated culture environment, some species may never be redescribed. New species must be described with more thorough documentation and voucher preparation. Even though the Code of Botanical Nomenclature (14) must be followed, it does not preclude more exhaustive measures to place the taxonomic literature on a firmer scientific foundation. This means that new germplasm should be established in culture as mycorrhizae and either maintained by the discoverer or by an established culture collection. If no other recourse is available, these fungi can by cryopreserved through facilities such as those of INVAM.

Systematic studies are in their infancy, with all of the attendant problems. First, and foremost, morphological characters must be redefined to reflect their origins, their boundary conditions, and their significance taxonomically and evolutionarily. Developmental comparisons are a basic necessity in species redescriptions and in reaching the ultimate goal of predicting diversity. When these results are coupled with cladistic analyses, robust decisions of homology are possible that define common ancestry. As discussed above, only phylogenetic relationships in glomalean fungi are capable of delimiting species and other taxa as natural groups.

Each branch in the cladogram and each major evolutionary lineage serves as a hypothesis which can be tested using all available characters. At other levels of organization within the fungal organism, characters must be discovered and incorporated into phylogenetic hypotheses. Then, the extent to which the organismal phenotype is linked to genetic and biochemical phenomena can be determined and form can be linked to function. In addition to conserved regions of the 18S rRNA gene, other genes or gene fragments must be sought out and sequenced. Less informative characters derived from DNA-DNA hybridization techniques or RFLP comparisons still provide important knowledge of genetic affinities and rates of molecular change. Number, stability, and distribution of characters generated by FAME pattern analysis will determine usefulness in phenetic or cladistic analyses. Immunological methods ultimately may be useful for diagnostic purposes among isolates of a species, but they have limited taxonomic value.

As taxonomic and systematic information becomes more organized and stable, biogeographic studies will provide meaningful information on patterns of species diversity and spore dispersal (16). If dispersal patterns can be correlated with geological phenomena (*e.g.*, continental drift), then placement of salient events in glomalean evolution in a temporal framework may be possible.

Many pressing questions remain unresolved, with the main hindrance being the inability to culture these fungal organisms independent of their host plants. If and when independent culture is achieved, then the question arises of the artificiality of conclusions based on fungi grown outside their natural environment. These problems are opposite sides of the same coin: To what extent is diversity in mycorrhizal associations causally linked to the fungus, the host, or to the interaction? This question, as it pertains to the fungus, can be answered by comparative studies examining the rate and magnitude of variation at the genic, and possibly even at the biochemical, level (*e.g.*, FAME analyses). As more molecular and biochemical markers are developed, genetic variation may be assessed both vertically (throughout generations of a single lineage) and horizontally (among different taxa). These studies are necessary if organismal biology is to be linked to physiological effects resulting from the mycorrhizal association.

Morphological evidence suggests that glomalean endomycorrhizal fungi are an evolutionary cul-de-sac (23,31), so that their uniqueness from the rest of the fungal kingdom is under-appreciated. However, only developmental, biochemical and genetic evidence will be able to link endomycorrhizal fungi with their nonmycorrhizal relatives. These kinds of studies also will help us to understand the nature and strength of epigenetic constraints during spore differentiation that account for the stability of taxonomic characters.

CONCLUSIONS

The current taxonomy of glomalean fungi has many problems. Fortunately, short-term solutions exist that will not impair research on functional aspects of mycorrhizal processes. Long-term solutions require systematic investigations that explain levels and kinds of taxonomic diversity according to pertinent biological processes operating within the fungal organisms. The current classification provides hypotheses of phylogeny based on changes in spore morphological characters. As characters are revised to reflect realities of spore differentiation processes, species will be redescribed and new species predicted based on existing patterns of diversity. This will benefit taxonomic purposes devoted to identification and also provide new and important information on mode of speciation. Epigenetic factors minimizing morphological variation have made taxonomic characters possible that define species and higher taxa, but they also have the more detrimental effect

of hiding genetic, physiological, and ecological
diversity so important in functional mycorrhizae. As
new characters are discovered at the genetic,
cellular, and ecological levels, the relationship
between molecular and organismal evolution will be
experimentally deduced. These kinds of diversity are
likely to differentiate individual populations, thus
providing a more realistic appraisal of organisms as
the functional units of comparison in these processes.
Most assuredly, the structure of classification will
change to accommodate new discoveries. Despite some
conflicts associated with instability, the final
product will help, rather than hinder, progress in
understanding all aspects of the mycorrhizal
symbiosis.

LITERATURE CITED

1. Almeida, R. T. 1989. Scientific names in the
    Endogonales, Zygomycetes. Mycotaxon 36: 147-159.
2. Almeida, R. T., and Schenck, N. C. 1991. A
    revision of the genus *Sclerocystis* (Glomaceae,
    Glomales). Mycologia 82: 703-714.
3. Berch S. M. 1987. Endogonaceae: Taxonomy,
    specificity, fossil record, phylogeny. Front.
    Appl. Microbiol. 2: 161-188.
4. Berch, S. M. 1988. Compilation of the
    Endogonaceae. Mycologue Publication, Waterloo,
    Ontario.
5. Berch, S. M., and Koske, R. E. 1986. *Glomus
    pansihalos*, a new species in the Endogonaceae,
    Zygomycetes. Mycologia 78: 832-836.
6. Berch, S. M., and Trappe, J. M. 1985. A new
    species of Endogonaceae, *Glomus hoi*. Mycologia
    77: 654-657.
7. Bhattarcharjee, M., Mukerji, K. G., Tewari, J.
    P., and Skoropad, W. P. 1982. Structure and
    hyperparasitism of a new species of *Gigaspora*.
    Trans. Brit. Mycol. Soc. 78: 184-188.
8. Cracraft, J. 1982. Nonequilibrium theory for
    the rate-control of speciation and extinction and
    the origin of macroevolutionary patterns. Syst.
    Zool. 31: 348-365.

9. Cronquist, A. 1988. The Evolution and Classification of Flowering Plants, Second Edition. The New York Botanical Garden, New York.

10. Cummings, B., and Wood, T. 1989. A simple and efficient method for isolating genomic DNA from endomycorrhizal spores. Gene Anal. Technol. 6: 89-92.

11. Franke, M. 1992. Relationships among species and isolates of arbuscular mycorrhizal fungi based on morphology, ontogeny and plant-fungus interaction. M.S. Thesis, West Virginia University, Morgantown, WV.

12. Gerdemann, J. W., and Trappe, J. M. 1974. The Endogonaceae in the Pacific Northwest. Mycologia Mem. 5: 1-76.

13. Gibson, J. L., Kimbrough, J. K., and Benny, G. L. 1986. Ultrastructural observations on Endogonaceae (Zygomycetes). II. Glaziellales ord. nov. and Glaziellaceae fam. nov.: New taxa based upon light and electron microscopic observations of *Glaziella aurantiaca*. Mycologia 78: 941-954.

14. Greuter, W. Burdet, H. M., Chaloner, W. T., Demoulin, V., Grolle, R., Hawksworth, D. L., Nicolson, D. H., Silva, P. C., Stafleu, F. A., Voss, E. G., and McNeil, J. 1988. International Code of Botanical Nomenclature. in: Regnum Vegatabile 118. Koeltz Scientific Books, Königstein. 328 pp.

15. Hall, I. R. 1984. Taxonomy of VA mycorrhizal fungi. Pages 57-94 in: VA Mycorrhiza, C. L. Powell, and D. J. Bagyaraj, eds., CRC Press, Boca Raton, FL.

16. Humphries, C. J., and Parenti, L. R. 1986. Cladistic Biogeography. Clarendon Press, Oxford.

17. Jabaji-Hare, S. 1988. Lipid and fatty acid profiles of some vesicular-arbuscular mycorrhizal fungi: contribution to taxonomy. Mycologia 80: 622-629.

18. Kluge, A. G., and Strauss, R. E. 1985. Ontogeny and systematics. Ann. Rev. Ecol. Syst. 16: 247-268.

19. Kohn L. M. 1992. Developing new characters for

fungal systematics: An experimental approach for determining the rank of resolution. Mycologia 84: 139-153.

20. Lewis, D. H. 1987. Evolutionary aspects of mutualistic associations between fungi and photosynthetic organisms. Pages 161-178 in: Evolutionary Biology of the Fungi, A. D. M. Rayner, C. M. Brasier, and D. Moore, eds., Cambridge University Press, New York, NY.

21. Lösel, D. M. 1988. Fungal lipids. Pages 699-805 in: Microbial Lipids, Vol. 1, C. Ratledge and S. G. Wilkinson, eds., Academic Press, London.

22. Malloch, D. 1987. The evolution of mycorrhizae. Can. J. Plant Pathol. 9:398-402.

23. Malloch, D., Pirozynski, K. A., and Raven, P. H. 1980. Ecological and evolutionary significance of mycorrhizal symbiosis in vascular plants (a review). Proc. Nat. acad. Sci. 77: 2113-2118.

24. Mayr, E. 1982. The Growth of Biological Thought: Diversity, Evolution, and Inheritance. Belknap Press, Cambridge, MA.

25. Mishler, B. D., and Brandon, R. N. 1987. Individuality, pluralism, and the phylogenetic species concept. Biol. Phil. 2: 397-414.

26. Mishler, B. D., and Donoghue, M. J. 1982. Species concepts: a case for pluralism. Syst. Zool. 31: 491-503.

27. Morton, J. B. 1985. Variation in mycorrhizal and spore morphology of *Glomus occultum* and *Glomus diaphanum* as influenced by plant host and soil environment. Mycologia 77: 192-204.

28. Morton. J. B. 1986. Three new species of *Acaulospora* (Endogonaceae) from high aluminum low pH soils in West Virginia. Mycologia 78: 641-648.

29. Morton, J. B. 1988. Taxonomy of VA mycorrhizal fungi: classification, nomenclature, and identification. Mycotaxon 32: 267-324.

30. Morton, J. B. 1990. Evolutionary relationships among arbuscular mycorrhizal fungi in the Endogonaceae. Mycologia 82: 192-207.

31. Morton, J. B. 1990. Species and clones of arbuscular mycorrhizal fungi (Glomales, Zygomycetes): Their role in macro-and

microevolutionary processes. Mycotaxon 37: 493-515.

32. Morton, J. B. 1993. Problems and solutions for the integration of glomalean taxonomy, systematic biology, and the study of endomycorrhizal phenomena. Mycorrhiza (in press).

33. Morton, J. B., and Benny, G. L. 1990. Revised classification of arbuscular mycorrhizal fungi (Zygomycetes): A new order, Glomales, two new suborders, Glomineae and Gigasporineae, and two new families, Acaulosporaceae and Gigasporaceae, with an emendation of Glomaceae. Mycotaxon 37: 471-491.

34. Morton, J. B., and Bentivenga, S. P. 1993. Levels of diversity in endomycorrhizal fungi (Glomales, Zygomycetes) and their role in defining taxonomic and nontaxonomic groups. Plant Soil (in press).

35. Morton, J., Franke, M., and Cloud, G. 1992. The nature of fungal species in Glomales (Zygomycetes). Pages 65-73 in: Mycorrhizas in Ecosystems, D. J. Read, D. H. Lewis, A. Fitter, and I. Alexander, eds., CAB International, University of Arizona Press, Tucson, AZ.

36. Mukerji, K. G., Bhattacharjee, M., and Tewari, J. P. 1983. New species of vesicular-arbuscular mycorrhizal fungi. Trans. Brit. Mycol. Soc. 81: 641-643.

37. Nicolson, T. H. 1975. Evolution of vesicular-arbuscular mycorrhizas. Pages 26-34 in: Endomycorrhizas, F. E. Sanders, B. Mosse, and P. B. Tinker, eds., Academic Press, London.

38. Patterson, C. 1982. Morphological characters and homology. Pages 21-74 in: Problems of Phylogenetic Reconstruction, eds., Joysey, K. A. and Friday, A. E. Academic Press, London.

39. Pirozynski, K. A., and Dalpé, Y. 1989. Geological history of the Glomaceae with particular reference to mycorrhizal symbiosis. Symbiosis 7: 1-36.

40. Pirozynski, K. A., and Malloch, D. W. 1975. The origin of land plants: A matter of mycotropism. BioSystems 6: 153-164.

41. Rieppel, O. C. 1988. Fundamentals of Comparative Biology. Birkhauser Verlag, Basel.
42. Rosendahl, S. 1989. Comparisons of spore cluster-forming *Glomus* species (Endogonaceae) based on morphological characteristics and isoenzyme banding patterns. Opera. Bot. 100: 215-223.
43. Sanders, I. R., Ravolanirina, F., Gianinazzi-Pearson, V., Gianinazzi, S., and Lemoine, M. C. 1992. Detection of specific antigens in the vesicular-arbuscular mycorrhizal fungi *Gigaspora margarita* and *Acaulospora laevis* using polyclonal antibodies to soluble spore fractions. Mycol. Res. 96: 477-480.
44. Simon, L., LaLonde, M., and Bruns, T. D. 1990. Specific amplification of 18S ribosomal genes from vesicular-arbuscular endomycorrhizal fungi colonizing roots. Appl. Environ. Microbiol. 58: 291-295.
45. Sneath, P. H., and Sokal, R. R. 1973. Numerical Taxonomy. W. H. Freeman and Co., San Francisco, CA.
46. Spain, J. L., Sieverding, E., and Schenck, N. C. 1989. *Gigaspora ramisporophora*: A new species with novel sporophores from Brazil. Mycotaxon 34: 667-677.
47. Stuessy, T. F. 1990. Plant Taxonomy: The Systematic Evaluation of Comparative Data. Columbia University Press, NY.
48. Stuessy, T. F. 1992. The systematics of arbuscular mycorrhizal fungi in relation to current approaches to biological classification. Mycorrhiza 1: 113-121.
49. Tommerup, I. C., and Sivasithamparam, K. 1990. Zygospores and asexual spores of *Gigaspora decipiens*, an arbuscular mycorrhizal fungus. Mycol. Res. 94: 897-900.
50. Trappe, J. M. 1982. Synoptic keys to the genera and species of Zyogmycetous mycorrhizal fungi. Phytopathology 72: 1102-1108.
51. Trappe, J. M., and Schenck, N.C. 1982. Taxonomy of the fungi forming endomycorrhizae. Pages 1-9 in: Methods and Principles of Mycorrhizal

Research, N. C. Schenck, ed., Am. Phytopathol. Soc., St. Paul, MN.

52. Van Valen, L. 1976. Ecological species, multispecies and oaks. Taxon 25: 233-239.

53. Wagner, G. P. 1989. The origin of morphological characters and the biological basis of homology. Evolution 43: 1157-1171.

54. Walker, C. 1983. Taxonomic concepts in the Endogonaceae: spore wall concepts in species descriptions. Mycotaxon 18: 443-455.

55. Walker, C. 1986. Taxonomic concepts in the Endogonaceae: II. A fifth morphological wall type in endogonaceous spores. Mycotaxon 25: 95-99.

56. Walker, C. 1987. Problems in taxonomy of mycorrhizal fungi. Pages 605-609 in: Les Mycorhizes: Physiologie et génétique, Gianiazzi-Pearson, V. and Gianinazzi, S. ed., INRA, Paris.

57. Walker, C. 1991. *Scutellospora* is *Scutellospora*. Mycotaxon 40: 141-143.

58. Walker, C., Reed, L. E., and Sanders, F. E. 1984. *Acaulospora nicolsonii*, a new endogonaceous species from Great Britain. Trans. Brit. Mycol. Soc. 83: 360-364.

59. Walker, C., and Sanders, F. E. 1986. Taxonomic concepts in the Endogonaceae: III. The separation of the genus *Scutellospora* gen. nov. from *Gigaspora* Gerd & Trappe. Mycotaxon 27: 169-182.

60. Walker, C., and Trappe, J. M. 1993. Names and epithets in the Glomales and Endogonales. Mycol. Res. 97:339-344.

61. Watrous, L. E., and Wheeler, Q. D. 1981. The out-group method of character analysis. Syst. Zool. 30: 1-11.

62. Weijman, A. C. M., and Meuzalaar, H. L. C. 1979. Biochemical contributions to the taxonomic status of the Endogonaceae. Can. J. Bot. 57: 284-291.

63. Wiley, E. O. 1981. Phylogenetics: The Theory and Practice of Phylogenetic Systematics. John Wiley, New York.

64. Wright, S. F., Morton, J. B., and Sworobuk, J. E. 1987. Identification of a vesicular-arbuscular mycorrhizal fungus by using monoclonal antibodies in an enzyme-linked immunosorbent assay. Appl. Environ. Microbiol. 53: 2222-2225.

MOLECULAR AND GENETIC APPROACHES TO UNDERSTANDING VARIABILITY IN MYCORRHIZAL FORMATION AND FUNCTIONING

Bradley R. Kropp and Anne J. Anderson
Utah State University
Logan, Utah

Sustainability in agriculture, forestry and range management requires balanced, functional microbial ecosystems. The association of plant roots with mycorrhizal fungi is a key factor in the belowground network essential to ecosystem function. These associations have been reported to benefit plants under conditions of nutritional and water stress and pathogen challenge. Molecular and genetic tools are, and will be used increasingly, to explore the structural and regulatory genes in both fungus and plant that permit mycorrhizal formation. Our understanding of the genetic loci that govern mycorrhizal formation will aid in our ability to manage ecosystems and manipulate the symbiosis to maintain sustainability (1,16,26,49,61,62, 67,87,88,93).

However, the mycorrhizal status in roots of plants is extremely varied. Mycorrhizal formation is associated with many diverse fungi, and the types of mycorrhizae that are formed differ in structure and efficacy (1,9,68). This review addresses features of two prevalent mycorrhizal classes: the ectomycorrhizae common with woody species, and thus of concern to forestry, and the endomycorrhizae more often associated with herbaceous plants with relevance to range, horticultural and agronomic plants. Recent reviews discuss the varied aspects of the ecology, biochemistry, physiology and molecular nature of mycorrhizal formation (1,10,11,30,43,79). Consequently, we highlight certain examples to address

how genetic and molecular studies in these areas can
be used to understand the basis of variability in
mycorrhizal formation, function and population
dynamics.

Variability in whether a plant will or will not
permit mycorrhizal formation occurs naturally, and
also as a result of laboratory manipulations with both
endo- and ectomycorrhizal fungi. As discussed by
Molina et al. (68), although only a small percentage
of plants are nonmycorrhizal, "the actual number is
certainly in the tens of thousands of species". With
the ectomycorrhizal fungi, some are generalists
whereas others are highly specific and form functional
mycorrhizae with only certain host species (68). The
endomycorrhizal fungi are unable to colonize certain
hosts although the majority of plants are compatible
(68). In the field, preferences between
endomycorrhizal fungi are seen for a host and
differences in functional efficiency are measured for
various host-fungal combinations (68,77). Virtually
nothing is known about the specific genes that
regulate mycorrhizal symbioses. However, it is
possible to use the variation that exists between
different hosts and fungal genotypes to gain clues
relevant to understanding the genetic regulation of
the symbiosis.

GENETICS OF MYCORRHIZAL INCOMPATIBILITY IN ECTOMYCORRHIZAL FUNGI: THE USE OF MONOKARYONS AND DIKARYONS

Ectomycorrhizal fungi offer a unique opportunity
to explore the role of specific nuclei and cytoplasmic
elements in mycorrhizal formation. The ectomycorrhizal basidiomycetes produce basidiospores that
are monokaryotic and germinate to produce haploid (or
monokaryotic) cultures. All of the fungi thus far
examined have a bifactorial (tetrapolar), multiallelic
mating system. This type of system involves two
mating type factors, usually designated as A and B,
each of which has multiple alleles which allow a
compatible mating to occur when they differ at each
factor (45,48). Because of this system, most

monokaryotic isolates of ectomycorrhizal
basidiomycetes are sexually compatible with a large
number of other monokaryons of the same species. A
number of papers describe work that uses monokaryotic
and reconstituted dikaryotic cultures of
ectomycorrhizal fungi to study the genetic variation
of characters thought to be important in the
functioning of the symbiosis (15,20,22,28,45,46,
47,48,50). However, in none of these studies is there
a correlation between mating type and ectomycorrhizal
ability. Therefore, it can be postulated that the
genes responsible for sexual incompatibility are not
linked with genes regulating the events involved in
ectomycorrhizal formation.
     Monokaryotic cultures of certain species do not
appear to be capable of ectomycorrhiza formation.
Ducamp et al. (22) reported that single-spore isolates
of *Suillus granulatus* (Fr.) Kuntze were incapable of
forming ectomycorrhizae while heterokaryotic cultures
rapidly formed normal associations with pine (*Pinus*
sp.) roots. Thus, we can infer that some species of
fungi require information in addition to that carried
in a haploid genome to be capable of ectomycorrhiza
formation and a functional symbiosis.
     In other studies (15,20), monokaryons of
Hebeloma cylindrosporum Romagnesi differed little from
dikaryons in their potential for ectomycorrhiza
formation. The ectomycorrhizae were identical to
those formed by dikaryons at both the macroscopic and
ultrastructural level. For example, ultrastructural
localization of acid phosphatase activity was
identical in ectomycorrhizae formed by both
monokaryotic and dikaryotic cultures of *H.
cylindrosporum*. Previous studies had shown the
monokaryons to vary greatly in their utilization of
nitrogen sources and soluble phosphate as well as in
their production of IAA (28,66,74,89).
     Studies of ectomycorrhiza formation by
monokaryotic progeny from a single sporocarp of
Laccaria bicolor (Maire) Orton showed that the
relative abilities of monokaryons to colonize pine
roots vary enormously (50,92). Some of the monokaryon
cultures were unable to form ectomycorrhizae even

though their hyphae directly contacted the root surface. Others readily formed a mantle and Hartig net when in contact with the roots of their host. The continuous nature of variability between haploid isolates suggests that ectomycorrhiza formation is polygenically controlled, perhaps in much the same way as radial growth rate is controlled in *Schizophyllum commune* Fr. (81). If this were the case, then we would expect to find both additive and nonadditive genetic variation affecting ectomycorrhiza formation to occur when different haploid cultures were paired.

This concept is supported by the finding (48) that the ectomycorrhiza-forming ability of dikaryons, resulting from crosses between selected sibling *L. bicolor* monokaryons, varied from 31 to 84 percent in colonization of available short roots of *Pinus strobus* L. In one array, where a single selected parental monokaryon was paired with four other monokaryons, the percent short roots colonized varied from 50-83 percent. When another parental monokaryon was paired with the same four monokaryons, root colonization only varied between 65-75 percent. Statistical analysis of 32 pairings between different parental monokaryons indicated that the total amount of heritable variation was between 43-71 percent. Roughly half the heritable genetic variation was additive, while the remainder was due to dominance at allelic and/or nonallelic sites. Rosado et al. (75) also found that variation among reconstituted dikaryotic cultures of *Pisolithus tinctorius* (Pers.) Coker & Couch to produce ectomycorrhizae was related more to nonadditive than additive interactions between parental monokaryons.

CYTOPLASMIC VERSUS NUCLEAR REGULATION OF THE SYMBIOSIS

In most cases the interaction between haploid genomes of ectomycorrhizal fungi that are aggressive and those that are nonmycorrhizal results in dikaryons that can form ectomycorrhizae. However, rare crosses between nonmycorrhizal and mycorrhizal monokaryons have resulted in dikaryons that are virtually unable to colonize the roots of their known hosts (50). This observation raises two different questions; the first being whether the nonmycorrhizal character is dominant in these instances and the second whether a cytoplasmic factor, such as a virus or a mitochondrial deletion, is involved in the regulation of ectomycorrhiza formation.

To address these questions, a nonmycorrhizal (M1) was crossed with a mycorrhizal (M2) monokaryon of *L. bicolor* to yield the dikaryon lacking in capacity for mycorrhiza formation (47). Monokaryotic cultures carrying M1 and M2 nuclei were recovered, from protoplasts made by enzymatic digestion of the mycelium of the nonmycorrhiza-forming dikaryon, and identified by their mating type genes (Figure 1). All of the regenerated monokaryotic cultures of the M1 type were incapable of ectomycorrhiza formation, whereas the regenerated M2 type monokaryons had ability to form ectomycorrhizae (Figure 1). When these derived monokaryons were again crossed, the resulting dikaryotic strains were at best very poor at forming mycorrhizae.

Nuclear migration in the basidiomycetes usually occurs without accompanying migration of cytoplasmic genetic determinants such as mitochondria. Although recombinant mitochondrial genomes have been isolated from the interface between paired monokaryotic cultures, this is thought to be relatively rare, and the nuclei from donor hyphae are found in the unmodified cytoplasmic background of the receiving mycelium (63,64). Much less is known about migration of other cytoplasmic elements (eg. viruses, etc.). Nonetheless, in the M1-M2 study, the reisolated M1 and M2 monokaryons would be contained in the identical cytoplasm because they were derived from the same dikaryon. Because the newly derived M1 and M2 monokaryons are different in their ability to form ectomycorrhizae, cytoplasmic factors are not important in determining ectomycorrhiza-forming potential in these crosses. Thus, a second implication of this study is that the nonmycorrhiza-forming character can be dominant when the haploid nuclei containing genes for mycorrhiza formation were directly confronted with one another in the same dikaryotic cell.

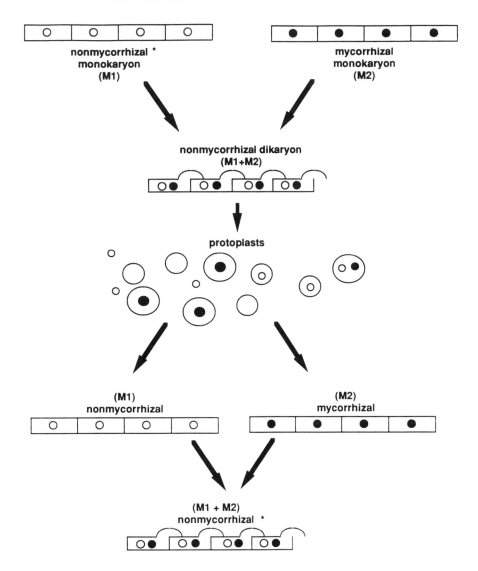

Figure 1. The recovery of both nonmycorrhizal and mycorrhizal monokaryons from the cytoplasm of a nonmycorrhizal dikaryon indicates that cytoplasmic factors have little control over mycorrhiza formation. The nonmycorrhizal character is dominant when mycorrhizal and nonmycorrhizal monokaryons are paired.

GENETICS OF FUNCTION RATHER THAN FORMATION

Most of the comparative studies with monokaryotic and dikaryotic cultures have considered either the formation of a mantle and Hartig net in short roots or single attributes implicated in ectomycorrhizal function, such as the production of IAA or phosphatases. Yet, little has been done to measure the effectiveness of the ectomycorrhizae that are formed.

Lamhamedi et al. (53) present data which indicate that *P. tinctorius*, requires the heterokaryotic state in order to fully express its' symbiotic ability. There was a much stronger correlation (r^2 = 0.515) between percent ectomycorrhizal development and total seedling dry weight when dikaryotic cultures were used than when monokaryotic cultures were involved (r^2 = 0.051). Lamhamedi and Fortin (52) also found that monokaryotic strains of *P. tinctorius* were completely unable to form rhizomorphs and that a heterokaryotic state was required for rhizomorphs to be formed. They reported that seedlings ectomycorrhizal with strains that produce rhizomorphs survived much longer under water stress than did seedlings ectomycorrhizal with strains that lacked the ability to produce rhizomorphs (53,54).

If one puts these findings of variable complexity in the genetics of ectomycorrhiza formation into an adaptive perspective, some sort of mechanism must exist to balance the interactions between ectomycorrhizal fungi and their hosts. There would be no benefit to the fungus or plant if the plant became totally nonmycorrhizal, and at the same time fungal symbionts that are too aggressive would likely become pathogenic on their host plant. Thus, it can be postulated that the overall effects of the additive and allelic and nonallelic interactions involved in ectomycorrhiza formation would balance the symbiosis. Polygenic interactions would tend to push the symbiotic balance towards the center (Figure 2), and would act to prevent both excessive and weak root colonization. However, polygonic inheritance per se is not an explanation for stability.

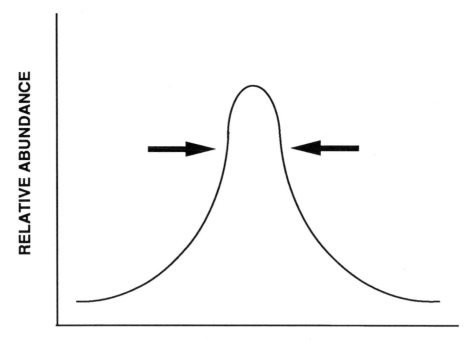

MYCORRHIZATION

Figure 2. Polygenic interactions may act to balance mycorrhization at a level where the fungus can benefit from the symbiosis without becoming parasitic on its host.

MOLECULAR TOOLS TO UNDERSTAND THE PHYSIOLOGY OF MYCORRHIZAL INCOMPATIBILITY

The complexity and polygenic character of the regulation of mycorrhizal formation becomes understandable when the variability is examined at the physiological level. The physiological interactions that are observed in both ectomycorrhiza and endomycorrhiza formation dictate that many genes in the fungus and in the plant are required. These genes regulate events at several stages of the interaction, initially in the rhizosphere and then at the root surface and finally within the root (3,10,44).

Signalling in Rhizosphere Events

Incompatibility is observed both before and after invasion of the host by the fungus (2,43,50). Certain monokaryotic and dikaryotic isolates of *L. bicolor* grow along the root surface but fail to invade (50). Similarly beet and other nonhost plants support rhizoplane growth of endomycorrhizal fungi, but no penetration structures form in contrast to the events on roots of a compatible host (2,6)

Are there signals required for such penetration structures to be produced, and are they lacking, inhibited or not recognized in the nonmycorrhizal plant? Attachment may be facilitated by specific recognition events. Lectins with different hapten specificities are associated with the hyphae of ectomycorrhizal *Lactarius* species. These lectins appear to recognize components on the host root hairs (34). Grafting experiments with nonhost and host plants suggest that a signal for incompatibility travels from shoots of lupin (*Lupinus albus* L.), a nonhost plant for endomycorrhizal fungi, to the roots of a host plant, pea (*Pisum sativum* L.) (31). Such signals may relate to the synthesis of secondary products by discrete cells in the plant roots. The recent review by Koide and Schreiner (43) includes excellent discussions of the potential of secondary compounds, such as the glucinolates, to inhibit mycorrhiza formation and govern host range. Also discussed is the concept that flavonoids stimulate the differentiation of hyphal structures of endo-mycorrhizal fungi needed to initiate successful penetration of the root. Stimulatory effects of different flavonoids are reported for several endomycorrhizal fungi (8,29,71,83,84). Increased concentrations of CO_2 that might be expected in the rhizosphere act as an additional stimulant (6,8).

Similar signalling systems occur with the root-associated bacteria, the pathogenic *Agrobacterium* and the symbiotic beneficial *Rhizobia* where flavonoids both stimulate and inhibit colonization (76). The detection of the flavonoids involves two component receptor-activator systems that enhance expression of bacterial genes concerned with the synthesis of machinery needed for colonization of the plant tissue. We await the studies to determine whether analogous detection systems exist in the mycorrhizal fungi. Genetic modifications of these systems could be one site at which incompatibility between a plant and a mycorrhizal fungus would be altered.

Signalling in Plant Interactions

Further genetic loci are likely involved in the penetration of the hyphae and the establishment of the symbiosis. Additional signals may be needed for formation of appressoria and the subsequent penetration hyphae with the mycorrhizal fungi. In this context, it is interesting that root factors stimulate the ATPase activity in hyphae from the germinating spores of an endomycorrhizal fungus (56). Also, an isoflavanoid has been identified in roots of mycorrhizal but not nonmycorrhizal soybean (*Glycine max* L.) (69).

The use of model gnotobiotic systems, such as endomycorrhizal formation in transformed roots (5,7,70), and plate, tube or pouch cultures of ectomycorrhizae (57) in which chemicals can be artificially applied may be useful in resolving the biochemistry and the genetics of the penetration process. Extraction of DNA from the fungus and from the plant and mRNAs from the responding tissues would permit cloning of mycorrhizal specific DNA's. Reprobing with *in situ* techniques would identify whether the gene is specific to the plant or fungus.

An approach of using mutant host plants affords additional understanding of the plant genes involved in mycorrhiza formation. Abnormal penetration structures are cited as one of the causes of incompatibility with nonmycorrhizal, chemically-derived mutants of plants that normally are endomycorrhizal. Abnormal appressoria form and the plant tissue rapidly responds with callose deposition which appears to limit further penetration (11,21,33). Such a deposition of callose is also an example of an induced resistance response to potential pathogenic challenges (35). It is interesting that some but not all of the *Myc*⁻ mutant plants are also resistant to

nodulation by *Rhizobium*, indicating some common
function of plant components(s) in these two types of
symbiosis (33). Comparative analysis with the
Rhizobium system, which is easier to work with both
genetically and biochemically than the mycorrhizal
symbiosis, thus is helping to identify specific plant
functions involved in endomycorrhiza formation.
Molecular techniques of hybridization at the DNA and
mRNA levels under conditions of differing stringency
using heterologous probes from plants and pathogenic
fungi will help determine how widespread is the
involvement of a particular gene product in
mycorrhizal development in different hosts.

Response of the fungus to the resistance
mechanisms of the plant is another site of interaction
where incompatibility is determined. Studies of
compatible systems suggests that plant defense
responses may be suppressed or expressed only on a
selective basis. Lambais and Mehdy (51) indicate
suppression of endochitinase, endoglucanase and
chalcone isomerase, and Blee and Anderson (unpublished
data) see no elevation of mRNA accumulations for
phenylalanine ammonia lyase and chalcone synthase, in
endomycorrhizal systems. However, expression of a
novel acidic chitinase in endomycorrhizal but not
pathogen-infected roots of tobacco (*Nicotiana tabacum*
L.) occurs (24). Allen et al. (2), in discussion of
the variety of ways in which nonmycorrhizal plants
respond to endomycorrhizal fungal challenge, describe
the intense interactions with roots of *Salsola kali* L.
and *Atriplex rosea* L. Allen et al. (2) liken the
response of browning, accumulation of fluorescent
compounds, which are likely to be phenolics, and root
necrosis to the hypersensitive response that is
commonly observed when plants are challenged by
incompatible pathogens. Genetic studies of this type
of incompatibility with bacterial plant pathogens
indicates that a battery of genes are required in the
hrp loci that are essential for both compatible and
incompatible interactions (91). The function of this
cluster in incompatibility is regulated by separate
specificity determining genes (41,91). The genes that
specify incompatibility, the avirulence genes, do so

by correspondence with genes in the host plant (41).
The same avirulence gene may specify resistance in a
cultivar and in a nonhost plant (42). Whether
analogues of *hrp* and avirulence genes exist to
determine mycorrhizal specificity is unknown.

FUNCTIONAL MYCORRHIZAE

Another measurable variable is the efficacy of
formed mycorrhizae. Efficacy may be determined by the
potential of the fungus to grow both within and
outside of the root. As discussed previously,
controlled crosses with *P. tinctorius* demonstrate that
extramatricial growth is genetically regulated (52).
Understanding the genetics of this trait may be
beneficial in engineering better mycorrhizal fungus
strains because production of extramatrical hypha is
associated with improved plant survival under drought
stress (52). RFLP analysis may help locate the loci
conditioning extramatrical growth. The use of chef
gels to separate large DNA fragments of chromosomal
size, or the products produced by the action of rare
cutting enzymes (36) followed by hybridization with
the RFLP, may permit allocation of the loci to a
particular chromosome.

The complexity and variable nature of internal
fungal growth patterns is extensively documented for
ecto- and endomycorrhizae by microscopy (9,39).
Current cytological methods to locate specific
cellular structures indicate how these are involved in
mycorrhiza formation. In a review by Bonfante-Fasolo
and Perotto (9), the results of cytology to probe the
subtle changes in the plant cell wall polysaccharide
and protein components associated with hyphal growth
and arbuscule formation in mycorrhizae are discussed.
One cytochemical method employs lectins to recognize
specific carbohydrate structures in the mycorrhizae.
A recent example demonstrated that Concanavalin A, a
lectin which recognizes α-linked glucose and mannose,
differentiates the interface regions for an aggressive
dikaryon and a weakly colonizing dikaryon derived from
crosses of the same monokaryons from the
ectomycorrhizal fungus *L. bicolor* (57,92). The

interface with the aggressive partner had greater
amounts of Concanavalin A reactive material (57). A
second technique uses the recognition potential of an
enzyme for its substrate to localize the substrate.
Analysis of cellulosic structures by cellobiose
recognition is one example (10). A third approach
uses both monoclonal and polyclonal antibodies derived
against the component to be studied. Antibodies have
been prepared against the fungal components and used
to distinguish isolates and detect the fungus within
the roots of plants (2,78). On the plant side, an
extensive array of monoclonal antibodies prepared
against arabinogalactan proteins and other plant wall
and plasmamembrane components (13,14), that have been
used for developmental analyses (12) and studies of
symbiosis with *Rhizobium* (76), are now being used to
examine the mycorrhizal roots (14,32,73). Again, as
in analysis of rhizobial symbiosis, current
cytological techniques coupled with the use of mutants
will be powerful in identifying key plant and fungal
players in mycorrhiza formation at both the
biochemical and genetic levels.

   Analysis by 2D gels of protein patterns in
extracts from mycorrhizal and nonmycorrhizal roots is
a direct method to detect proteins novel to the
mycorrhizal root. Mycorrhiza-distinct proteins are
reported for several endomycorrhizal and
ectomycorrhizal systems (23,38,72) as discussed by
Bonfante-Fasolo and Perrotti (10). Comparison of 2D
gels from different hosts challenged with the same
fungus may help determine whether similar plant
proteins are involved. This approach has been
successful in surveys of the role of pathogenesis-
related proteins in induced resistance responses in
plants (35,55). Molecular techniques initiated by
sequencing the N termini of these proteins can permit
the synthesis of corresponding nucleic acid probes and
the generation of complementary antibodies for use in
in situ analysis. Of interest is the finding that
some of the mycorrhiza-specific proteins in an
endomycorrhiza are in common with plant proteins
produced by *Rhizobium* nodulation (94). Perhaps these
root cell proteins guide the invading hypha and help

limit potential plant defense responses against the mycorrhizal fungus.

Pinpointing the changes in plant proteins during mycorrhiza formation will permit construction of transgenic plants altered in these properties. Expression of transformed genes to up regulate the amount of product, or antisense RNA to reduce the product, could be employed. The ability of these transgenics to become mycorrhizal should permit further understanding of specific plant structures essential for mycorrhiza formation.

Genetically altered mycorrhizal fungi may be also be constructed. Molecular biology tools for fungal analysis are available and are being sharpened as discussed by Metzenberg (65). Vectors, promoter libraries and selectable markers such as resistance to benomyl, bleomycin and hygromycin are being used in transformation of mycorrhizal fungi (4,60,93). Transformation, now possible by several methods including protoplast uptake, microinjection and biolistics, will permit discrete DNA manipulations (93). The construction of mutants to reduce or overexpress activities is also feasible. Using this approach Durand et al. (25) have derived overproducers of IAA to initiate studies of the role of this hormone in mycorrhiza formation with the ectomycorrhizal fungus *H. cylindrosporum*. Perhaps from understanding the complexity of the physiology of mycorrhizae we can account for the genetic behavior and explain the subtle differences between various isolates and hosts in mycorrhizal efficacy.

GENETIC APPROACHES TO POPULATION STUDIES IN THE FIELD

In addition to their value in understanding the basic molecular interaction between plant and mycorrhizal fungi, molecular and genetic techniques may also generate novel information on these fungi under field conditions. Why are certain isolates dominant? What governs the change in populations with cropping systems (40)? What are the mechanisms that give a particular fungus superior mycorrhizal potential so that it becomes a dominant clone? What

are the mechanisms of competition between the fungi
(19)? Are antibiotics, and degradative extracellular
enzymes involved, as they appear to be, in competition
between other soil microbes and the fungal pathogens
(19)?

Many species of fungi exist in nature as mycelial
clones composed of anastomosing hyphae which make up a
single genetic unit often referred to as genets. One
of the most striking examples, is the finding of a
single clone of a nonmycorrhizal basidiomycete,
Armillaria gallica Marx, Muller & Romagnesi,
(*Armillaria bulbosa* (Borla) Kile & Wattling),
estimated to be about 1,500 years old that occupies at
least 15 hectares (86). Even more striking is the
genetic stability of this clone. By analyzing RAPD
products and restriction fragments from nuclear DNA,
Smith et al. (86) found no evidence of genetic
variation nor indication for intraclonal variation in
mitochondrial DNA (85).

Using vegetative incompatibility as a means to
identify genets of the ectomycorrhizal fungus *Suillus
bovinus* (L. Fr.) young Scotch pine (*Pinus sylvestris*
L.) stands between 10-20 years old, had between 700-
900 genets per hectare averaging between 0.7-2 M in
size (17,18). Older stands had fewer genets (25-500
per hectare) that averaged 30 m in size). They also
found a pattern of decreasing fruiting per unit area
as stand age increased. These findings imply
basidiospores are important in the initial
establishment of ectomycorrhizal populations, but
decrease in importance as the population matures.
Another implication is certain genets gradually
dominate the site. Dahlberg (17) estimated that the
median life expectancy of a given genet is 35 years.
It is not known whether ectomycorrhizal fungal genets
can become as large or remain as genetically stable as
those of *A. gallica*. Neither is it known what
characteristics would allow one genet to dominate
while the others die out. In fact, it is not even
certain that the genets do die out. Since the
ectomycorrhizal fungus occurs primarily in the soil
and on the roots of its host, the presence or absence
of fruiting bodies may not correlate with the presence

of an active ectomycorrhizal fungus. Certain genets
may stop fruiting as the stand matures, yet their
mycelium might still be actively associated with the
tree roots.

Although vegetative incompatibility and isozymes
can be used to study populations of ectomycorrhizal
fungi (80), pure cultures or carpophores are required.
The development of the polymerase chain reaction (PCR)
offers a breakthrough in dealing with samples from the
field. Gardes et al. (27) have amplified both nuclear
and mitochondrial DNA directly from ectomycorrhizae
using primers that discriminate between plant and
fungal DNA. Their work also demonstrates that it is
feasible to identify ectomycorrhizal basidiomycetes
using RFLPs generated from ribosomal DNA amplified
directly from root samples. Similar studies have
demonstrated variation in specific DNA sequences
between isolates of *Cenococcum geophilum* Fr. (58), and
Hebeloma (59). Henrion et al. (37) also report
success in identifying ectomycorrhizal fungi using
restriction fragments from PCR amplified ribosomal
DNA. Simon et al. (81) have made primers that permit
the specific amplification of ribosomal DNA of VAM
fungi from mycorrhizal roots. Thus, PCR based
methodology might also be useful in studying the
colonization patterns and genetics of VAM fungal
populations (87).

Most of the work to date on fingerprinting
mycorrhizal fungi has been done using primers which
amplify the internal transcribed spacer region of
ribosomal genes. This region is roughly 600 base
pairs long and is variable enough to allow
identification of different fungal genera and species
(27,37). However, it is doubtful that it is variable
enough to allow discrimination between different
genets of a single species. Very little work has been
done to determine whether there are other DNA regions
which have the combined properties of enough
variability to allow discrimination between different
genets along with primer sites allowing the selective
amplification of fungal DNA. For example, the
intergenic spacer regions between the ribosomal repeat
units or even genes coding for certain enzymes may

turn out to be useful for identifying individual fungal strains or genets. RAPD products (90) may also be useful in studies of fungal population genetics.

Using these newly developed techniques it should be possible to identify not only fungal species, but also genets from mycorrhizal roots and soil mats. It should also be possible to follow the progress of a transgenic strain of a mycorrhizal fungus introduced into the field. The tools may aid researchers in documenting mycorrhizal succession and the composition of mycorrhizal communities.

CONCLUSIONS

Mycorrhizal research is entering a challenging and exciting period when molecular and genetic tools can be used synergistically. The development of techniques to permit studies of the mycorrhizal fungi which are difficult at best to culture will expand our horizons of the value and the functioning of the below-ground root-fungal symbiosis.

LITERATURE CITED

1.   Allen, M.F. 1992. Mycorrhizal Functioning. An Integrative Plant Fungal Process. Chapman and Hall, New York, NY, 534 pp.

2.   Allen, M.F., Clouse, S.D., Weinbaum, B.S., Jeakins, S., Friese, C.F., and Allen, E.B. 1992. Mycorrhizae and the integration of scales: From molecules to ecosystems. Pages 488-515 in: Mycorrhizal Functioning. An Integrative Plant Fungal Process. M.F. Allen, ed., Chapman and Hall, New York, NY.

3.   Anderson, A.J. 1992. The influence of the plant root on mycorrhizal formation. Pages 37-64 in: Mycorrhizal Functioning. An Integrative Plant Fungal Process. M. F. Allen, ed., Chapman and Hall, New York, NY.

4.   Barrett, V., Dixon, R.K., and Lemke, P.A. 1990. Genetic transformation of a mycorrhizal fungus. Applied Micro. & Biotech. 33:313-316.

5. Becard, G., and Fortin, J.A. 1988. Early events of vesicular-arbuscular mycorrhiza formation on Ri T-DNA transformed roots. New Phytol. 108:211-18.

6. Becard, G., and Piche, Y. 1989. Fungal growth stimulation by CO_2 and root exudates in vesicular-arbuscular mycorrhizal symbiosis. Appl. Environ. Microbiol. 55(9):2320-2325.

7. Becard, G., and Piche, Y. 1990. Physiological factors determining vesicular-arbuscular mycorrhizal formation in host and nonhost Ri T-DNA transformed roots. Can. J. Bot. 68:1260-1264.

8. Becard, G., Douds, D.D., and Pfeffer, P.E. 1992. Extensive in vitro hyphal growth of vesicular-arbuscular mycorrhizal fungi in the presence of CO_2 and flavonols. Appl. Environ. Micro. 58:821-825.

9. Bledsoe, C.S. 1992. Physiological ecology of ectomycorrhizae: implications for field application. Pages 424-437 in: Mycorrhizal Functioning. An Integrative Plant Fungal Process. M. F. Allen, ed., Chapman and Hall, New York, NY.

10. Bonfante-Fasolo, P., and Perotto, S. 1991. Plants and endomycorrhizal fungi: The cellular and molecular basis of their interaction. Pages 445-470 in: Molecular Signals in Plant-Microbe Communications, Verma, D.P.S.

11. Bonfante-Fasolo, P., Perotto, S., and Peretto, R. 1992. Cell surface interactions in endomycorrhizal symbiosis. in: Perspective in Plant Cell Recognition, J. Callow, and J. Green, eds., SEB Seminars, Cambridge University Press.

12. Bonfante-Fasolo, P., Tamagnone, L., Peretto, R., Esquerre-Tugaye, M.T., Mazau, D., Mosiniak, M., and Vian, B. 1991. Immunocytochemical location of hydroxyproline rich glycoproteins at the interface between a mycorrhizal fungus and its host plants. Protoplasma. 165:127-138.

13. Bonfante-Fasolo, P., Vian, B., Perotto, S., Faccio, A., and Knox, J.P. 1990. Cellulose and pectin localization in roots of mycorrhizal

Allium porrum: Labelling continuity between host cell wall and interfacial material. Planta 180:537-547.

14. Brewin, N.J., Robertson, J.G., Wood, E.A., Wells, B., Larkins, A.P., Galfre, G., and Butcher, G.W. 1985. Monoclonal antibodies to antigens in the peribacteroid membrane from *Rhizobium*-induced root nodules of pea cross-react with plasma membranes and Golgi bodies. The EMBO Journal 4:605-611.

15. Bruchet, G., Debaud, J.C., and Gay, G. 1986. Genetic variations in the physiology of *Hebeloma*. Pages 121-131 in: Physiological and genetical aspects of mycorrhizae Proc. 1st European Symposium on Mycorrhizae, V. Gianinazzi-Pearson and S. Gianinazzi, eds., INRA, Paris.

16. Crook, G. 1988. Using mycorrhized seedlings: a private sector viewpoint - CIP Inc. Pages 35-37 in: Proceedings of the Canadian Workshop on Mycorrhizae in Forestry, May 1-4, 1988, M. Lalonde and Y. Piche, eds., Centre de recherche en biologie forestière, Faculté de foresterie et de géodésie, Université Laval, Sainte-Foy (Québec).

17. Dahlberg, A. 1991. Ectomycorrhiza in coniferous forest: structure and dynamics of populations and communities. Sveriges Lantbruksuniversitet. Uppsala. Swedish University of Agricultural Sciences, Department of Forest Mycology and Pathology.

18. Dahlberg, A., and Stenlid, J. 1990. Population structure and dynamics in *Suillus bovinus* as indicated by spatial distribution of fungal clones. New Phytol. 115:487-493.

19. Deacon, J.W., and Fleming, L.V. 1992. Interactions of ectomycorrhizal fungi. Pages 249-300 in: Mycorrhizal Functioning. An Integrative Plant Fungal Process. M. F. Allen, ed., Chapman and Hall, New York, NY.

20. Debaud, J.D., Gay, G., Prevost, A., Lei, J., and Dexheimer, J. 1988. Ectomycorrhizal ability of gentically different homokaryotic and dikaryotic

mycelia of *Hebeloma cylindrosporum*. New Phytol.
108:323-328.

21. Duc, G., Trouvelot, A., Gianinazzi-Pearson, V.,
     and Gianinazzi, S. 1989. First report of
     nonmycorrhizal plant mutants (Myc⁻) obtained in
     pea (*Pisum sativum* L.) and fababean (*Vicia faba*
     L.), Plant Sci., 60:215.

22. Ducamp, M., Poitou, N., and Olivier, J.M. 1986.
     Comparaison cytologique et biochimique entre
     culture monospores et boutures du carpophores
     chez *Suillus granulatus* (Fr. ex L.) Kuntze.
     Pages 575-579 in: Physiological and genetical
     aspects of Mycorrhizae. Proc. 1st European
     Symposium on Mycorrhizae, V. Gianinazzi-Pearson
     and S. Gianinazzi, eds., INRA, Paris.

23. Dumas, Gaudat-E., Tahiri-Alaoui, A., Gianinazzi,
     S., and Gianinazzi-Pearson, V. 1990.
     Observations on modifications in gene expression
     with VA endomycorrhiza development in tobacco:
     qualitative and quantitative changes in protein
     profiles. Page 153 in: Endocytobiology IV. P.
     Nardon, V. Gianinazzi-Pearson, A. M. Grenier, L.
     Margulis, and D. C. Smith, eds., INRA, Paris.

24. Dumas, Gaudat-E., Furlan, V., Grenier, J.,
     Asselin, A. 1992. New acidic chitinase
     isoforms induced in tobacco roots by vesicular
     arbuscular mycorrhizal fungi. Mycorrhiza
     1:133-136.

25. Durand, N., Debaud, J.C., Casselton, L.A., and
     Gay, G. 1992. Isolation and preliminary
     characterization of 5-fluoroindole-resistant and
     IAA-overproducer mutants of the ectomycorrhizal
     fungus *Hebeloma cylindrosporum* Romagnesi. New
     Phytol. 121:545-553.

26. Fagel, R. 1980. Mycorrhizae and nutrient
     cycling in natural forest ecosystems. New
     Phytol. 86:199-212.

27. Gardes, M., White, T.J., Fortin, J.A., Bruns,
     T.D., and Taylor, J.W. 1990. Identification of
     indigenous and introduced symbiotic fungi in
     ectomycorrhizae by amplification of nuclear and
     mitochondrial ribosomal DNA. Can. J. Bot.
     69:180-190.

28. Gay, G., and Debaud, J.C. 1987. Genetic study on indole-3-acetic acid production by ectomycorrhizal Hebeloma species: inter- and intraspecific variability in homo- and dikaryotic mycelia. Appl. Microbiol. Biotechnol. 26:141-146.

29. Gianinazzi-Pearson, V., Branzanti, B., Gianinazzi, S. 1989. In vitro enhancement of spore germination and early hyphal growth of a vesicular-arbuscular mycorrhizal fungus by host root exudates and plant flavonoids. Symbiosis 7:243-255.

30. Gianinazzi-Pearson, V. and Gianinazzi, S. 1989. Cellular and genetical aspects of interactions between hosts and fungal symbionts in mycorrhizae. Genome 31:336.

31. Gianinazzi-Pearson, V., and Gianinazzi, S. 1992. Influence of intergeneric grafts between host and non-host legumes on formation of vesicular-arbuscular mycorrhiza. New Phytol. 120:505-508.

32. Gianinazzi-Pearson, V., Gianinazzi, S., and Brewin, N.J. 1990. Immunocytochemical localisation of antigenic sites in the perisymbiotic membrane of vesicular-arbuscular endomycorrhiza using monoclonal antibodies reacting against the peribacteroid membrane of nodules. Page 153 in: Endocytobiology IV. P. Nardon, V. Gianinazzi-Pearson, A. M. Grenier, L. Margulis, and D. C. Smith, eds., INRA, Paris.

33. Gianinazzi-Pearson, G.S., Guillemin, J.P., Trouvelot, A., and Duc, G. 1991. Genetic and cellular analysis of resistance to vesicular-arbuscular (VA) mycorrhizal fungi in pea mutants. Page 336 in: Advances in Molecular Genetics of Plant-Microbe Interactions. H. Hennecke and P. S. Verma, eds., Kluwer Academic Publishers, Netherlands.

34. Giollant, M., Guillot, J., Damez, M., Dusser, M., Didier, P., and Didier, E. 1993. Characterization of a lectin from *Lactarius deterrimus*. Plant Physiol. 101:513-522.

35. Graham, T.L., Graham, M.Y. 1991. Cellular

coordination of molecular responses in plant
defense. Molecular Plant-Microbe Interactions
5:415-422.

36. Grothues, D., Tummler, B. 1991. New approaches
in genome analysis by pulsed-field gel
electrophoresis: application to the analysis of
Pseudomonas species. Molecular Microbiology
5:2763-2776.

37. Henrion, B., Tacon, F.L., and Martin, F. 1992.
Rapid identification of genetic variation of
ectomycorrhizal fungi by amplification of
ribosomal RNA genes. New Phytol. 289-298.

38. Hilbert, J.L. and Martin, F.M. 1988. Regulation
of gene expression in Ectomycorrizae. I.
Protein changes and the presence of
ectomycorrhiza specific polypeptides in the
Pisolithus-Eucalyptus symbiosis. New Phytol.
110:339-346.

39. Jacquelinet-Jeanmougin, I., Gianinazzi-Pearson,
V., and Gianinazzi, S. 1987. Endomycorrhizas
in the gentianaceae. II. Ultrastructural
aspects of symbiont relationships in *Gentiana
lutea* L. Symbiosis 3:269-285.

40. Johnson, N.C., Pfleger, F.L., Crookston, R.K.,
Simmons, S.R., and Copeland, P.J. 1991.
Vesicular-arbuscular mycorrhizas respond to corn
and soybean cropping history. New Phytol.
117:657-663.

41. Keen, N.T., and Buzzell, R.I. 1991. New
disease resistance genes in soybean against
Pseudomonas syringae pv. *glycinea*: Evidence
that one of them interacts with a bacterial
elicitor. Theor. Appl. Genet. 81:133-138.

42. Kobayashi, D.Y., Tamaki, S.J., and Keen, N.T.
1989. Cloned avirulence genes from the tomato
pathogen *Pseudomonas syringae* pv. *tomato* confer
cultivar specificity on soybean. Proc. Natl.
Acad. Sci. USA 86:157-161.

43. Koide, R.T., and Schreiner, R.P. 1992.
Regulation of the vesicular-arbuscular
mycorrhizal symbiosis. Annu. Rev. Plant
Physiol. Plant Mol. Biol. 43:557-81.

44. Koske, R.E., and Gemma, J.N. 1992. Fungal

reactions to plants prior to mycorrhizal formation. Pages 3-36 in: Mycorrhizal Functioning. An Integrative Plant Fungal Process. M. F. Allen, ed., Chapman and Hall, New York, NY.

45. Kropp, B.R., 1989. Variation in acid phosphatase activity among progeny from controlled crosses in the ectomycorrhizal fungus *Laccaria bicolor*. Can. J. Bot. 68:864-866.
46. Kropp, B.R. 1990. Variable interactions between nonmycorrhizal and ectomycorrhizal strains of the basidiomycete *Laccaria bicolor*. Mycological Research 94:412-415.
47. Kropp, B.R. 1992. A genetic analysis of variation in ectomycorrhizal colonization by *Laccaria bicolor*. Phytopathology 82:1171.
48. Kropp, B.R., and Fortin, J.A. 1987. The incompatibility system and relative ectomycorrhizal performance of monokaryons and reconstituted dikaryons of *Laccaria bicolor*. Can. J. Bot. 66:289-294.
49. Kropp, B.R., and Langlois, C. 1990. Ectomycorrhizae in reforestation. Can. J. For. Res. 20:438-451.
50. Kropp, B.R., McAfee, B.J., and Fortin, J.A. 1987. Variable loss of ectomycorrhizal ability in monokaryotic and dikaryotic cultures of *Laccaria bicolor*. Can. J. Bot. 65:500-504.
51. Lambais, M.R., and Mehdy, M.C. 1993. Suppression of endochitinase, β-1, 3-endoglucanase, and chalcone isomerase expression in bean vesicular-arbuscular mycorrhizal roots under different soil phosphate conditions. Molecular Plant-Microbe Interactions 6:75-83.
52. Lamhamedi, M.S., and Fortin, J.A. 1991. Genetic variations of ectomycorrhizal fungi: extramatrical phase of *Pisolithus* sp. Can. J. Bot. 69:1927-1934.
53. Lamhamedi, M.S., Fortin, J.A., and Bernier, P.Y. 1991. La génétique de *Pisolithus* sp.: une approche de biotechnologie forestière pour une meilleure survie des plants en conditions de sécheresse. Sécheresse 2:251-258.

54. Lamhamedi, M.S., Fortin, J.A., Kope, H.H., and
     Kropp, B.R. 1990. Genetic variation in
     ectomycorrhiza formation by *Pisolithus arhizus*
     on *Pinus pinaster* and *Pinus baksiana*. New
     Phytol. 115:689-697.
55. Legrand, M., Kauffmann, S., Geoffroy, P., and
     Fritig, B. 1987. Biological function of
     pathogenesis-related proteins: Four tobacco
     pathogenesis-related proteins are chitinases.
     Proc. Natl. Acad. Sci. USA 84:6750-6754.
56. Lei, J., Becard, G., Catford, J.G., and Piche,
     Y. 1991. Root factors stimulate ^{32}P uptake and
     plasmalemma ATPase activity in vesicular-
     arbuscular mycorrhizal fungus, *Gigaspora
     margarita*. New Phytol. 118:289-294.
57. Lei, J., Wong, K.Y., and Piche, Y. 1991.
     Cellular concanavalin A-binding sites during
     early fractions between *Pinus banksiana* and two
     closely related types of the ectomycorrhizal
     basidiomycete *Laccaria*. Mycological Research
     95:357-363.
58. LoBuglio, K.F., Rogers, S.O., and Wang, C.J.K.
     1991. Variation in ribosomal DNA among isolates
     of the mycorrhizal fungus *Cenococcum geophilum*.
     Can. J. Bot. 69:2331-2343.
59. Marmeisse, R., Debaud, J.C., and Casselton,
     L.A. 1992. DNA probes for species and strain
     identification in the ectomycorrhizal fungus
     Hebeloma. Mycological Research 96:96-100.
60. Marmeisse, R., Gay, G., Debaud, J.C., Casselton,
     L.A. 1992. Genetic transformation of the
     symbiotic basidiomycete fungus *Hebeloma
     cylindrosporium*. Current Genetics. 22:41-45.
61. Marx, D.H. 1980. Ectomycorrhizal fungus
     inoculations: a tool for improving forestation.
     Pages 13-71 in: Tropical mycorrhiza research.
     P. Mikola, ed., Clarendon Press, Oxford.
62. Marx, D.H., and Cordell, C.E. 1988. Specific
     ectomycorrhizae improve reforestation and
     reclamation in the eastern United States. Pages
     75-86 in: Proceedings of the Canadian Workshop
     on Mycorrhizae in Forestry, May 1-4, 1988, M.
     Lalonde and Y. Piche, eds., Centre de recherche

en biologie forestière, Faculté de foresterie et de géodésie, Université Laval, Sainte-Foy (Québec).

63. May, G. 1988. Somatic incompatibility and individualism in the coprophilous basidiomycete, *Coprinus cinereus*. Trans. Br. Mycol. Soc. 91:443-451.

64. May, G., and Taylor, J.W. 1988. Patterns of mating and mitochondrial DNA inheritance in the agaric basidiomycete *Coprinus cinereus*. Genetics 118:213-220.

65. Metzenberg, R.I. 1991. Benefactors' lecture: The impact of molecular biology on mycology. Mycological Research 95:9-13.

66. Meyselle, J.P., Gax, G., and Debaud, J.C. 1986. Intraspecific variability in an ectomycorrhizal fungus; *Hebeloma cylindrosporum* 3. Soluble phosphate utilization by sib-monokaryons and wild dikaryons. Pages 597-604 in: Physiological and genetical aspects of mycorrhizae. Proc. 1st European Symposium Mycorrhizae, Dijon, 1985, V. Gianinazzi-Pearson and S. Gianinazzi, eds., INRA, Paris.

67. Mikola, P. 1973. Application of mycorrhizal symbiosis in forestry practice. Pages 383-406 in: Ectomycorrhizae, their ecology and physiology. G.C. Marks and T.T. Kozlowski, eds., Academic Press, New York.

68. Molina, R., Massicotte, H., and Trappe, J.M. 1992. Specificity phenomena in mycorrhizal symbioses: Community-ecological consequences and practical implications. Pages 357-423 in: Mycorrhizal Functioning. An Integrative Plant Fungal Process. M. F. Allen, ed., Chapman and Hall, New York, NY.

69. Morandi, D., and LeQuere, J.L. 1991. Influence of nitrogen on accumulation of sosojagol (a newly detected coumestan in soybean) and associated isoflavonoids in roots and nodules of mycorrhizal and nonmycorrhizal soybean. New Phytol. 117:75-79.

70. Mosse, B., and Hepper, C.M. 1975. Vesicular-arbuscular mycorrhizal infections in root organ

cultures. Physiol. Plant Pathol. 5:215-223.

71. Nair, M.G., Safir, G.R., Siqueira, J.O. 1991.
     Isolation and identification of vesicular-
     arbuscular mycorrhiza-stimulatory compounds from
     clover (*Trifolium repens*) roots. Appl. Environ.
     Microbiol. 57(2):434-39.

72. Pacovsky, R.S. 1989. Carbohydrate, protein and
     amino acid status of Glycine-*Glomus*-
     Bradyrhizobium symbioses. Physiol. Plant.
     75:346-353.

73. Perotto, S., Vandenbosch, K.A., Brewin, N.J.,
     Faccio, A., Knox, J.P., and Bonfante-Fasolo, P.
     1990. Modifications of the host cell wall
     during root colonization by Rhizobium and VAM
     fungi. Page 114 in: Endocyto-biology IV, P.
     Nardon, V. Gianinazzi-Pearson, A. M. Grenier, M.
     Margulis, and D. C. Smith, eds., INRA, Paris.

74. Plassard, C., Scheromm, P. Porcher, P.,
     Mourrais, D., Bousquet, N., Tillard, P.,
     Labarere, J., and Salsac, P. 1987. Nitrate
     reductase and phosphatase activities in
     ectomycorrhizal isolates. Page 258 in:
     Mycorrhizae in the next decade: practical
     applications and research priorities. Proc. 7th
     N. Amer. Conf. on Mycorrhizae. Gainesville,
     1987. D.M. Sylvia. L.L. Hung, and J.H. Graham,
     eds., IFAS, University of Florida, Gainesville.

75. Rosado, S.C.S., Kropp, B.R., and Piché, Y.
     1993. Genetics of ectomycorrhizal symbiosis:
     2. Fungal variability and heritability of
     ectomycorrhizal characteristics. New Phytol.
     (In press)

76. Sanchez, F., Padilla, J.E., Perez, H., and Lara,
     M. 1991. Control of nodulin genes in root-
     nodule development and metabolism. Annu. Rev.
     Plant Physiol. Plant Mol. Biol. 42:507-28.

77. Sanders, I.R., and Fitter, A.H. 1992. Evidence
     of differential response between host-fungus
     combinations of vesicular-arbuscular mycorrhizas
     from a crossland. Mycological Research 96:415-
     419.

78. Sanders, I.R., Ravolanirina, F., Gianinazzi-
     Pearson, V., Gianinazzi, S., and Lemoine, M.C.

1992. Detection of specific antigens in the vesicular-arbuscular mycorrhizal fungi *Gigaspora margarita* and *Acaulospora laevis* using polyclonal antibodies to soluble spore fractions. Mycological Research 96:477-480.

79. Schwab, S.M., Menge, J.A., and Tinker, P.B. 1991. Regulation of nutrient transfer between host and fungus in vesicular-arbuscular mycorrhizas. New Phytol. 117:387-398.

80. Sen, R. 1989. Intraspecific variation in two species of *Suillus* from Scots pine (*Pinus sylvestris* L.) forests based on somatic incompatibility and isozyme analyses. New Phytol. 114:607-616.

81. Simchen, G., and Jinks, J.L. 1964. The determination of dikaryotic growth rate in the basidiomycete *Schizophyllum commune*: A biometrical analysis. Heredity 19:629-649.

82. Simon, L., Lalonde, M., and Bruns, T.D. 1992. Specific amplification of 18S fungal ribosomal genes from vesicular-arbuscular endomycorrhizal fungi colonizing roots. Appl. Environ. Microbiol. 58:291-295.

83. Siqueira, J.O., Safir, G.R., and Nair, M.G. 1990. Stimulation of vesicular-arbuscular mycorrhiza formation and growth of white clover by flavonoid compounds. New Phytol. 118:87-93.

84. Siqueira, J.O., Safir, G.R., Nair, M.G. 1991. Stimulation of vesicular-arbuscular mycorrhiza formation and growth of white clover by flavonoid compounds. New Phytol. 118:87-93.

85. Smith, M.L., Duchesne, L.C., Bruhn, J.N., and Anderson, J.B. 1990. Mitochondrial genetics in a natural population of the plant pathogen *Armillaria*. Genetics 126:575-582.

86. Smith, M.L., Bruhn, J.N., and Anderson, J.B. 1992. The fungus *Armillaria bulbosa* is among the largest and oldest living organisms. Nature:356:428-431.

87. Tommerup, I.C. 1992. The role of mycorrhiza in plant populations and communities. Mycorrhiza 1:123-125.

88. Vogt, K.A., Grier, C.G., Meier, C.E., and

Edmonds, R.L. 1982. Mycorrhizal role in net primary production and nutrient cycling in *Abies amabilis* ecosystems in western Washington. Ecology 63:370-380.

89.   Wagner, F., Gay, G., and Debaud, J.C. 1986. Intraspecific variability in an ectomycorrhizal fungus: *Hebeloma cylindrosporum* 2. Ammonium and nitrate utilization by sib-monokaryons and wild dikaryons. Pages 589-596 in: Physiological and Genetical aspects of Mycorrhizae. Proc. 1st European Symposium on Mycorrhizae, Dijon, 1985. V. Gianinazzi-Pearson and S. Gianinazzi, eds., INRA, Paris.

90.   Williams, J.G.K., Kubelik, A.R., Livak, K.J., Rafalski, J.A., and Tingey, S.V. 1990. DNA polymorphisms amplified by arbitrary primers are useful as genetic markers. Nucleic Acids Research 18:6531-6535.

91.   Willis, D.K., Rich, J.J., and Hrabak, E.M. 1991. *hrp* genes of phytopathogenic bacteria. Molecular Plant-Microbe Interactions 4:132-138.

92.   Wong, K.Y., Piché, Y., Montpetit, D., and Kropp, B.R. 1988. Differences in the colonization of *Pinus banksiana* roots by sib-monokaryotic and dikaryotic strains of ectomycorrhizal *Laccaria bicolor*. Can. J. Bot. 67:1717-1726.

93.   Wood, T., and Cummings, B. 1992. Biotechnology and the future of VAM commercialization. Pages 468-487 in: Mycorrhizal Functioning. An Integrative Plant Fungal Process. M. F. Allen, ed., Chapman and Hall, New York, NY.

94.   Wyss, P., Mellor, R.B., and Wiemken, A. 1990. Vesicular-arbuscular mycorrhizas of wild type soybean and non-nodulating mutants with *Glomus mosseae* contain symbiosis specific polypeptides (mycorrhizins), immunologically cross-reactive with nodulins. Planta 182:22.

GENERAL SUMMARY

R. G. Linderman
USDA-ARS Horticultural Crops Research Laboratory
Corvallis, Oregon

and

F. L. Pfleger
Department of Plant Pathology
St. Paul, Minnesota

Nearly all terrestrial plants on Earth have or could have mycorrhizae of one type or another. The major types of mycorrhizae of importance in agriculture and forestry are VA mycorrhizae (VAM) and ectomycorrhizae. The evidence is very strong that plants evolved with mycorrhizae, and in natural ecosystems are highly dependent on them for their contributions to growth, health, and survival. Mycorrhizae are involved in many fundamental plant processes because they link plants and soil and induce changes in the host plant physiology. For plant roots to have mycorrhizae is as normal and essential to the plant as for plant leaves to have chlorophyll. If our crop plants, other than those that are nonmycotrophic, lack mycorrhizae, it is probably because the detrimental effects agricultural and forestry practices have had on these fungi. In some cases, those practices cause shifts in the mycorrhizal fungal species composition of a soil, resulting in less effective mycorrhizal fungus strains or species.

Plant growth and health in all ecosystems depends on the maintenance of optimum physical structure, and chemical and biological balances in the soil. Soil systems can be drastically disturbed by cultivation,

337

compaction, crop removal, fertilizer and pesticide applications, irrigation, and environmental fluctuations. As a result, the balances required for optimum plant growth and health can be altered. Degraded soil systems lack the physical, chemical, and biological components they once had, and farmers can generally only compensate by adding back chemical substances.

Plant growth and health is supported in many ways by rhizosphere microbes, and key among those microbes are mycorrhizal fungi. Rhizosphere microorganisms influence many chemical reactions by way of their metabolites, and mycorrhizae play a crucial role in facilitating both microbial and plant functions as mediators of exchanges between them. Mycorrhizae improve the health and development of plants by enhancing nutrition, modifying physiological functions of plants, reducing plant response to environmental stresses, and modifying the chemistry and biology of the rhizosphere in ways that alter nutrient cycling and suppress activity of root pathogens. The extraradical phase of mycorrhizae extends out in soil and generates significant changes in soil aggregation, organic matter accumulation, and microbial activity in the hyphasphere soil; all these changes improve the structure and "health" of the soil. If any type of microorganism could induce and orchestrate interactions and functions of the soil in relation to the growth and health of plants, it is mycorrhizal fungi. The following discussion summarizes the ways mycorrhizae contribute to plant health.

Disease Suppression

When mycorrhizae form, the host plant physiology changes significantly in response to the presence of the fungal symbiont within root tissues. The mycorrhizal fungus can produce, or induce the host plant to produce, altered levels of chemical constituents, including phytohormones, isoflavones, etc. that can influence plant growth and function. A significant change in mycorrhizal plants is the altered root exudation of materials into the rhizosphere soil that cause the selection of a new microbial equilibrium in the "mycorrhizosphere". Some of those materials can directly or indirectly inhibit root pathogens, i.e. production of oxalic acid that directly inhibits the Fusarium root rot pathogen of conifers or alters the pH such that the pathogen is suppressed. Other changes occur in the root tissues that involve the production of inhibitory phenolic compounds that inhibit the pathogen. With ectomycorrhizae, the fungal mantle also acts as a mechanical barrier to ingress by root pathogens. With VAM fungi, changes in host constitutes, like increased levels of phytoalexin-like isoflavonoid substances or specific amino acids like arginine, or induced morphological changes in root tissues can block pathogen ingress or modify its development in the host tissues. A less explored but highly likely mechanism of suppression of pathogens is the selective increase in populations of antagonists in the mycorrhizosphere as a result of altered patterns of exudation. Such microbes and/or changes in host plant physiology could account for the reduced incidence or severity of root diseases caused by fungal or nematode pathogens. Antagonistic populations or microbes have not been analyzed in most studies, so this mechanism is still speculative. Diseases caused by bacterial pathogens are decreased by VAM, while diseases caused by viruses or many foliage pathogens are often increased because they thrive in or on plants made healthier by VAM. VAM and ectomycorrhizae are known to reduce plant response to abiotic stresses, such as drought and nutrient deficiency, that may predispose plants to biotic infections by opportunistic pathogens. At present, it is not possible to predict the usefulness of mycorrhizae in disease management, but they should be viewed as having high potential in integrated pest management programs.

Land Reclamation

Restoration or reclamation of disturbed sites, largely due to mining operations, with plants representing the natural biodiversity of the area or of the site before disturbance, is a major challenge. Disturbance greatly alters the chemical, physical and biological characteristics of the planting medium. As a result, attempted revegetation frequently fails. Studies indicate that a major reason for plant failure is the lack of mycorrhizal fungi that aid in the establishment and growth of introduced plants. Populations of indigenous mycorrhizal fungi in the topsoil may have declined greatly during the storage period of the mining operation. The planting medium of the reclamation site, such as coal, oil, shale, iron tailing, etc. differ based on the parent geologic material and processes used for extraction. Also, the climatic conditions associated with mining activities differs from one geographic region to another and influence revegetation strategies. Evidence indicates that mycorrhizae can greatly aid in the revegetation process by improving the nutrient-acquiring capacity of roots, and by increasing tolerance of plants to soil drought and toxicities. The challenge is to select the best mycorrhizal fungi and compatible plants for each site, and introduce them together. Because VAM fungi are not available in large, commercial quantities, introducing pre-inoculated plants as island sources of VAM inoculum can aid in plant succession on the reclaimed sites. Inoculum of ectomycorrhizal fungi is commercially available and plants have been successfully inoculated in the nurseries before transplanting into the disturbed sites. Growth responses of ectomycorrhizal transplants in these harsh sites has been very dramatic in numerous geographic locations throughout the world.

Cultural Practices

The exploitation of mycorrhizae in land reclamation, agriculture, or forestry is often influenced by the cultural practices employed. For example, it is now known that crop rotation, minimum tillage and the use of cover crops improve crop vigor by positively affecting VAM formation. Other practices that could encourage the VAM symbiosis are breeding for increased VAM responsiveness to reduce plant dependence on high fertilizer input, modification in crop sequence, and reduced fertilizer input. Pesticide usage may inhibit VAM formation, but many pesticides have no effect and some encourage VAM, so careful selection of pesticides may be needed to encourage VAM. Pesticides applied in tree nurseries appear to have little effect on ectomycorrhizae except for the inhibitory effect of Bayleton used to control fusiform rust. It is critical, however, to select the proper fungal symbiont for the tree species and the planting site. At present, most seedling nurseries fumigate the soil before planting, a practice that reduces pathogens and weeds, but also reduces competition with the introduced fungal symbiont by weedy or indigenous fungi that may form less effective mycorrhizae.

Air Pollution

The effects of air pollutants, such as ozone, carbon dioxide, or acid rain on plant growth and health has been studied extensively, but only recently has part of the damage been connected to effects on mycorrhizae. Even when damage to the foliage is not detectable, mycorrhiza formation and function can be impaired due to impacts of pollutants on carbon partitioning, rhizosphere characteristics, and nutrient balances within plants and between plants and soil. These effects can have major implications on plant ecology, but more quantitative descriptions of relationships between plants, stresses, and mycorrhizae are needed before the role of mycorrhizae can be assessed.

Biogeochemistry

Mycorrhizae link the biotic and geochemical portions of the ecosystems, but it is extremely difficult to measure their impact on ecosystem responses. Mycorrhizae are believed to contribute to biogeochemical cycling of nutrients more than by simply providing a greater hyphal surface area for scavenging nutrient elements that may be relatively immobile in soil or in short supply. They retain nutrients by altering their concentration ratios in vegetation and decreasing their mobility by retaining a greater proportion of nutrient ions in biomass. The combination of these functions means that the system relies more on mineralization processes of nutrient acquisition than on degrading parent soil materials. In addition, mycorrhizae make major contributions to nutrient availability by increased accumulation and protection of organic matter in soil aggregates that they induce. In addition, soil aggregates greatly influence soil structure that can significantly influence aeration, percolation, and soil binding that reduces erosion.

There is some evidence that ectomycorrhizae have greater selective nutrient uptake than roots, and possibly greater uptake efficiency. They may also mediate weathering of minerals and organic materials through production of enzymes, organic acids, and siderophores. Evidence for similar activity by VAM fungi is lacking, largely because they have not been cultured.

Genetic Diversity: Genetics and Systematics

Mycorrhizal fungi are groups of soil fungi that have great genetic diversity. The VAM fungi are grouped in the Glomales, based on differences in spore morphological characters, and are represented by relatively few taxa. Systematic investigations that explain levels and kinds of taxonomic diversity as determined by pertinent biological processes functioning within fungal organisms are needed to provide a long-term solution to the taxonomy of this group. However, though VAM fungi lack large morphological differences, epigenetic factors hide genetic, physiological, and ecological differences that greatly affect their function. Ectomycorrhizal fungal diversity is represented by both great numbers of taxa and great variation in strains within taxa. Undoubtedly epigenetic factors also affect their function in the ecosystem. VAM fungal genetics is largely unknown because the fungi have not been cultured. However, the future analysis of their multi-nucleate spores could lead to the discovery of genetic mechanisms that account for diversity in function. With ectomycorrhizal fungi, studies on segregated homokaryon cultures have demonstrated

genetic control of mycorrhiza formation and function.
Molecular genetic studies have also indicated that
gene products are involved in recognition phenomena
between fungal and plant symbionts. Host products
such as some isoflavonoids have been shown to
stimulate VAM formation. DNA probes have been
developed for specific mycorrhizal fungi that will aid
in taxonomy as well as in the identification of
unknown mycorrhizal symbionts *in situ* in soil or
roots. Research on genetic diversity of mycorrhizal
fungi is entering an exciting period.

Practical Application Technology

How well developed is our technology for
application and exploitation of the benefits of
mycorrhizae in agriculture, forestry, and land
reclamation? For VAM fungi, a major obstacle has been
unavailability of large quantities of commercial
inoculum. The fact that VAM fungi are obligate
symbionts necessitates that inoculum be generated on
living plant roots. The pot culture method is useful
for experimental purposes, but other methods have now
been developed that result in greater quantities of
fungal inoculum; for example, the nutrient film and
aeroponic methods. Commercial inoculum has not been
readily available, although sources in Canada and
Japan have been identified. Commercial inoculum of
ectomycorrhizal fungi is available as vegetative
inoculum on a vermiculite carrier, as alginate pellets
containing mycelium, and as spores of some hypogeous
fungi. Inoculation technology is driven by the form
of inoculum, with most being delivered as soil amended
inoculum (both VAM and ectomycorrhizae) except for
basidospores of ectomycorrhizal fungi that can be
introduced to seedlings through irrigation systems.
In reclamation of mine sites, plants inoculated in the
nursery become the inoculum source after outplanting.
Similarly, reforestation depends exclusively on
outplanting seedlings pre-inoculated in the nursery
with fungi selected for their ease of handling and for
their capacities to influence survival and growth of
seedlings under stress conditions. Many outplanting
studies are currently underway throughout the world,

and success is largely a function of the compatibility of the selected fungal symbiont with the host species and the biotic and abiotic conditions of ,the outplanting site.

Managing indigenous inoculum of VAM fungi in agricultural systems appears to be possible, based on the discoveries that other crop management practices, such as crop rotation and fallow, significantly affect population shifts in VAM fungal species, some of which are more effective in plant growth enhancement than others. Furthermore, some sequences lead to buildup of chemicals in the soil that are toxic to VAM fungi. Other management practices, such as tillage, may greatly disturb populations of VAM fungi, leading to significantly decreased inoculum potential. On the other hand, isoflavonoid compounds have been identified that appear to enhance VAM formation that can contribute to crop yield increase.

The belief that mycorrhizae greatly influence plant health has been supported by the many examples discussed in this book. Agricultural and forestry practices may have decreased the potential for mycorrhizae to be effective due to population reductions, reduced biodiversity, or shifts in species composition in soil. The benefits of mycorrhizae in suppression of plant diseases; aiding in the reclamation of mine spoils; maintenance of nutrient cycling systems, structure and stability of soils; reducing the needs for excessive use of fertilizers and pesticides; stabilizing ecosystems exposed to air pollutants; and maintaining the biodiversity of all ecosystems, make them key to maintaining healthy plants on the planet Earth.